Studies in Modern Chemistry

Advanced courses in chemistry are changing rapidly in both structure and content. The changes have led to a demand for up-to-date books that present recent developments clearly and concisely. This series is meant to provide advanced students with books that will bridge the gap between the standard textbook and the research paper. The books should also be useful to a chemist who requires a survey of current work outside his own field of research. Mathematical treatment has been kept as simple as is consistent with clear understanding of the subject.

Careful selection of authors actively engaged in research in each field, together with the guidance of four experienced editors, has ensured that each book ideally suits the needs of persons seeking a comprehensible and modern treatment of rapidly developing areas of chemistry.

William C. Agosta, The Rockefeller University

R. S. Nyholm, FRS, University College London

Consulting Editors

Academic editor for this volume

T. C. Waddington, University of Durham

Studies in Modern Chemistry

R. L. M. Allen
Colour Chemistry

R. B. Cundall and A. Gilbert
Photochemistry

T. L. Gilchrist and C. W. Rees
Carbenes, Nitrenes, and Arynes

S. F. A. Kettle
Coordination Compounds

Ruth M. Lynden-Bell and Robin K. Harris
Nuclear Magnetic Resonance Spectroscopy

M. F. R. Mulcahy
Gas Kinetics

E. S. Swinbourne
Analysis of Kinetic Data

T. C. Waddington
Non-aqueous Solvents

K. Wade
Electron Deficient Compounds

Electron Deficient Compounds

K. Wade
University of Durham

Nelson

Thomas Nelson and Sons Ltd
36 Park Street London W1Y 4DE
PO Box 18123 Nairobi Kenya

Thomas Nelson (Australia) Ltd
597 Little Collins Street Melbourne 3000

Thomas Nelson and Sons (Canada) Ltd
81 Curlew Drive Don Mills Ontario

Thomas Nelson (Nigeria) Ltd
PO Box 336 Apapa Lagos

Thomas Nelson and Sons (South Africa) (Proprietary) Ltd
51 Commissioner Street Johannesburg

First published in Great Britain 1971

Copyright © K. Wade 1971

Softcover reprint of the hardcover 1st edition 1971

Illustrations by Colin Rattray

17 761707 1 (Boards)
17 771706 8 (Paper)

ISBN 978-1-4684-6056-8   ISBN 978-1-4684-6054-4 (eBook)
DOI 10.1007/978-1-4684-6054-4

Filmset by Keyspools Ltd, Golborne, Lancashire

C. Tinling & Co. Ltd, London and Prescot

Contents

A*

Preface

This book is about compounds such as the boron hydrides and associated metal hydrides and alkyls which acquired the label 'electron deficient' when they were thought to contain too few valence electrons to hold together. Though they are now recognized as containing the numbers of bonding electrons appropriate for their structures, the term 'electron deficient' is still commonly applied to many substances that contain too few valence electrons to provide a pair for every pair of atoms close enough to be regarded as covalently bonded.

The study of such substances has contributed much to chemistry. Techniques for the vacuum manipulation of volatile substances were devised specifically for their study; developments in valence theory resulted from considerations of their bonding; and the reactivity of several (for example, diborane and complex metal hydrides, lithium and aluminium alkyls) has made them valuable reagents.

The purpose of this book is to provide an introduction to the chemistry of these fascinating compounds. The experimental and spectroscopic methods by which they can be studied are outlined, the various types of structure they adopt are described and profusely illustrated, and the relative merits of extended valence bond and simple molecular orbital treatments of their bonding are discussed, with as liberal use of diagrams and as limited recourse to the Greek alphabet as possible. A recurring theme is the importance attached to considerations of molecular symmetry. Their reactions are treated in sufficient detail to show whether these reflect any deficiency of electrons. The occurrence of 'electron deficient' bonding in systems other than those commonly dubbed 'electron deficient' is also briefly noted.

The treatment is considered appropriate for the final year undergraduate. Where it is felt that the reader may be unfamiliar with the type of information provided by a particular spectroscopic method, explanatory comments are made. Reference is made to review articles that provide suitable further reading, and to original papers where these are particularly significant or provide leads into recent work. The literature is covered to mid-1970. Problems (with answers or guides thereto) are appended where appropriate.

I am grateful to Professor T. C. Waddington for many helpful discussions, and to Messrs J. B. Farmer and R. Snaith for assistance with literature searches.

K. Wade

Introduction

'There are no such things as electron deficient compounds, only theory deficient chemists'—R. E. Rundle.

This book is about compounds with too few valence electrons to provide a pair for every pair of atoms close enough to be regarded as covalently bonded.

In order to appreciate how such substances acquired the label 'electron deficient', it is necessary to consider the terms in which the electronic theory of valency was first formulated. When G. N. Lewis wrote his remarkably perceptive paper on the subject in 1916, generalizing from the compounds then known he concluded that a covalent bond was a link localized between two atoms which involved the sharing between those two atoms of a pair of electrons. The sharing of a pair of electrons between just two atoms accordingly came to be regarded as the fundamental unit of covalent bonding, indeed the minimum requirement for normal bonds, and it remains the 'single bond' of today; the two-centre two-electron bond is the unit relative to which bond orders are still expressed. Here we shall be concerned with compounds containing bonds that are formally of order less than one, compounds that contain too few valence electrons to make all their two-centre bonds electron-pair bonds.

Such compounds are frequently recognizable as electron deficient from their molecular formulae, in which n atoms are held together by fewer than $2(n-1)$ electrons; a minimum of $(n-1)$ two-centre links is required to hold n atoms together whatever the molecular structure happens to be. In the case of the classic example of electron deficiency, diborane, B_2H_6, for example, the minimum number of links needed to hold together its eight atoms would be seven, so even if its molecular

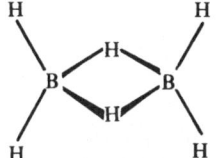

Fig. 1–1   Structure of diborane

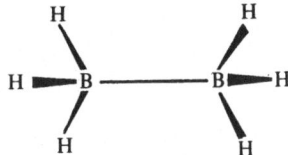

Fig. 1–2   Hypothetical ethane-like structure for diborane

structure (actually that shown in Fig. 1–1) had been like that of ethane, as in Fig. 1–2, the compound would still have been electron deficient, having fewer than 14 valence electrons. It was this feature of diborane and of the higher boron hydrides such as tetraborane B_4H_{10}, the pentaboranes B_5H_9 and B_5H_{11}, and decaborane $B_{10}H_{14}$ that showed them all to be electron deficient as soon as their molecular formulae were established in A. Stock's classic researches between 1910 and 1930, long before the details of their structures were known. The associated metal alkyls, such as dimeric trimethylaluminium, $[(CH_3)_3Al]_2$ and tetrameric methyllithium, $(CH_3Li)_4$, polymeric metal hydrides such as aluminium hydride $(AlH_3)_n$, and covalent borohydrides like the beryllium and aluminium compounds $Be(BH_4)_2$ and $Al(BH_4)_3$ are further examples of substances containing fewer than $2(n-1)$ valence electrons to hold n atoms together.

Certain other substances are less obviously electron deficient, but are seen to be so when their molecular structures are known and the number of two-centre links is found to exceed, sometimes considerably, the minimum number formally required to hold their atoms together. The borane anions, $B_mH_m^{2-}$, and isoelectronic carboranes, $B_{m-2}C_2H_m$, are species of this type. They contain $2(2m+1)$ valence electrons to hold $2m$ atoms together, four more than the minimum number needed to ensure that each atom is linked to at least one other. The actual structures of these substances are based on polyhedral cages of boron atoms, with in most cases five- or six-coordinate boron atoms, and therefore many more two-centre links than in a hypothetical (electron sufficient) chain structure.

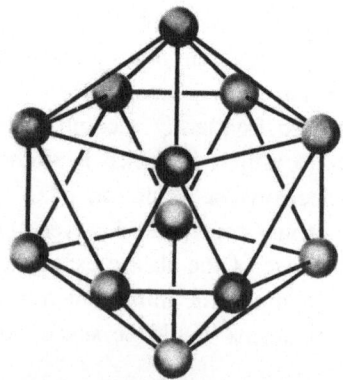

Fig. 1–3 Icosahedral structure of the $B_{12}H_{12}^{2-}$ anion; the single terminal hydrogen attached to each boron is omitted

The anion $B_{12}H_{12}^{2-}$, for example, has the icosahedral structure shown in Fig. 1–3, in which all the boron atoms are six-coordinate, and there are 32 two-centre bonds to be accounted for by a total of 50 bonding electrons.

An unusually high coordination number—the number of neighbouring atoms to which a particular atom is considered to be directly bonded—is a characteristic of certain of the atoms in most electron deficient molecules. Thus, diborane adopts the structure shown in Fig. 1–1, with the boron atoms four coordinate and two of the hydrogens two coordinate, rather than dissociate into two molecules BH_3 with trigonal planar structures and coordination numbers three and one for the boron and hydrogens respectively. The two coordinate hydrogens of diborane, and six coordinate borons of the $B_{12}H_{12}^{2-}$ anion, reveal another feature of electron deficient compounds, that is that as a general rule (to which there are but few exceptions) the coordination number of one or more of their constituent atoms exceeds the number of atomic orbitals in the valence shells of those atoms. Boron, with one s and three p atomic orbitals in its valence shell, the second principal quantum shell, frequently is found to have a coordination number greater than four. Hydrogen, with only its K shell $1s$ orbital, is at least two-coordinate when it acts as a bridging atom. These high coordination numbers underline the fact that electron deficient compounds are compounds that occupy the borderline between covalent and ionic bonding, or between covalent and metallic bonding. Ionic substances have structures in which the cations and anions are packed together as closely as possible, bearing in mind their relative sizes and proportions, and the number of near-neighbour contacts between cations and anions often exceeds considerably the number of electron pairs the anion would have to contribute to account for the structure in terms of localized two-centre electron pair bonds. For example, in the cubic lattice of sodium chloride, in which both sodium and chloride ions are six coordinate, each chloride ion would require a further two pairs of electrons in its valence shell if a pair were to be allocated to each point of contact between chloride and sodium ions. Again, metals adopt structures in which there are many more points of contact between adjacent metal atoms than there are pairs of electrons, and their bonding is best dealt with in terms of the band theory, with the bonding electrons able to move throughout the lattice accommodated in incompletely filled energy bands rather than in localized orbitals.

Appropriately then for compounds straddling the covalent–ionic–metallic borderlines, electron deficient compounds are formed principally by such elements as lithium, beryllium, magnesium, boron, and aluminium, which can contribute fewer valence electrons than they possess low energy atomic orbitals, which are sufficiently electropositive to form polar bonds to hydrogen or carbon with the latter elements at the negative end of the bond, but which are not electropositive enough, or are too strongly polarizing, to form essentially ionic bonds to these elements.

It has already been indicated that the boron hydrides, volatile metallic borohydrides, polymeric non-ionic metal hydrides, and associated metal alkyls are all conventionally regarded as electron deficient, as are the more recently discovered carboranes. Further examples are the so-called 'non-classical' or bridged structures believed to be adopted by certain carbonium ions, typified by the protonated cyclopropane structure (Fig. 1–4) attributed to the propyl cation to account for certain of its reactions. It is the chemistry of such systems that is outlined in the follow-

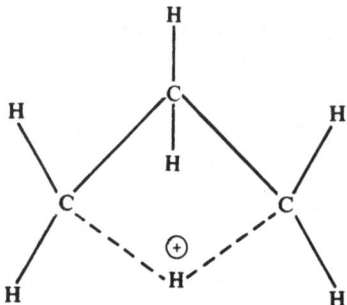

Fig. 1-4 Protonated cyclopropane structure possible for the propyl carbonium ion

ing pages. The probable occurrence of bonds of low order in some less obviously electron deficient species is also noted. Strictly, using the criteria of electron deficiency outlined above, one should also classify as electron deficient a number of transition metal complexes of aromatic ring systems such as dibenzene chromium, $(C_6H_6)_2Cr$, and ferrocene, $(C_5H_5)_2Fe$, which have sandwich structures (ferrocene is shown in Fig. 1–5) with coordination numbers for the metal of 12 and 10 respectively,

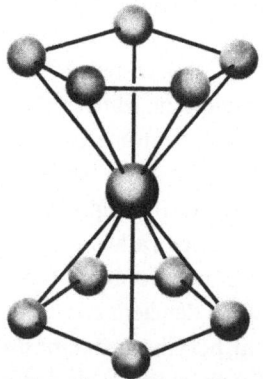

Fig. 1-5   Sandwich structure of ferrocene, $(\pi\text{-}C_5H_5)_2Fe$; O = iron; o = carbon; the single terminal hydrogen attached to each carbon is omitted

but only 18 electrons in the valence shell of the metal in each case. However, they are not normally described as electron deficient, and their chemistry is best discussed alongside that of other transition metal–hydrocarbon species that are not electron deficient, so reference is made to them here only so far as is judged necessary to put in perspective the chemistry of transition metal complexes of carboranes.

The study of electron deficient compounds has contributed much to chemistry. The experimental methods which had to be devised by Stock and his coworkers to allow the isolation and characterization of the boron hydrides have facilitated the study of many other reactive and/or volatile substances. Unravelling the valence problems posed by their structures has provided a tremendous stimulus to the development of valence theory and provided a striking vindication of molecular orbital theory. Some compounds, such as the lithium alkyls and diborane, have become useful laboratory reagents in preparative organic chemistry; others like the aluminium alkyls, are widely used as components of industrially important catalyst systems. It is the object of this book to provide an introduction to the chemistry of these fascinating compounds, with particular reference to those features that distinguish them from 'normal' compounds with at least two electrons per two-centre bond, so justifying their separate classification. The appropriateness of the label 'electron *deficient*' will also be considered, bearing in mind that this term is also used in organic chemistry to refer to atoms at the positive ends of polar bonds, atoms which act as sites for nucleophilic attack by electron-rich species, and also bearing in mind that a substance truly deficient in electrons might be expected to function as an electron (or electron-pair) acceptor, i.e. as an oxidizing agent or Lewis acid. The label 'electron deficient' was, after all, originally acquired when these compounds appeared to contain too few electrons to hold together. As indicated by the quotation at the start of this introductory chapter, the deficiency turned out to be in our understanding of chemical bonding rather than in the number of electrons they contain, which is just that appropriate for the structures they in fact adopt.

2 Structure and bonding of diborane

2–1 Introduction

Diborane played a key role in the development of our understanding of the bonding in electron deficient compounds. Although it was first isolated and characterized by Stock in 1912,[1] and recognized from its molecular formula as posing something of a valence problem, the distribution of its valence electrons in the molecule could not of course be determined until the positions of the atomic nuclei were known. The elucidation of its structure proved no easy task, and 30 years were to elapse before spectroscopic techniques were developed sufficiently for convincing evidence to be obtained that its structure was the bridged structure shown in Fig. 1–1, first suggested in 1921 by Dilthey,[2] rather than the ethane-like alternative shown in Fig. 1–2. The steps by which it was finally possible to distinguish between these structures reveal both the power and limitations of various methods of structure determination, and are briefly outlined in the following section.

2–2 Elucidation of the structure of diborane

Stock established the formula B_2H_6 for diborane by elemental analysis, and by calculation of the molecular weight from vapour density measurements. As diborane is very volatile (b.p. $-92.5°$, m.p. $-165°$), its molecules clearly have little affinity for their neighbours. An essentially ionic formulation such as $BH_2^+BH_4^-$ could therefore be discounted early, and the problem resolved itself into discrimination between such hypothetical covalent structures as those shown in Figs 1–1 and 1–2.

An early (1925) **X-ray diffraction** study of the solid did little more than roughly locate the boron atoms, which were found to be some 180 pm apart.[3] It is not surprising that the hydrogen atoms were not located, because the extent to which X-rays are diffracted is a function of the electron density in the matter through which they pass, and anyway the sample was in the form of a crystalline mass rather than a single crystal. Although hydrogen atoms do not scatter electron-beams much either, **electron-diffraction** studies on gases can be a much more sensitive method than X-ray crystallography for the location of hydrogen atoms in simple molecules, affording information about internuclear distances, whether bonded or non-bonded. Discriminating between the structures shown in

Figs 1–1 and 1–2 requires a precise knowledge of internuclear distances involving hydrogen, however, and the precision of the method, at the time that it was first applied to diborane in 1942, was poor, leading the researchers to conclude that diborane probably had a structure like ethane (Fig. 1–2).

Apart from the internuclear distances of the two structures (Figs 1–1 and 1–2), another important respect in which these models differ is in their **symmetry**. The various symmetry elements of the two models are illustrated in Figs 2–1 and 2–2. Because of the marked differences between

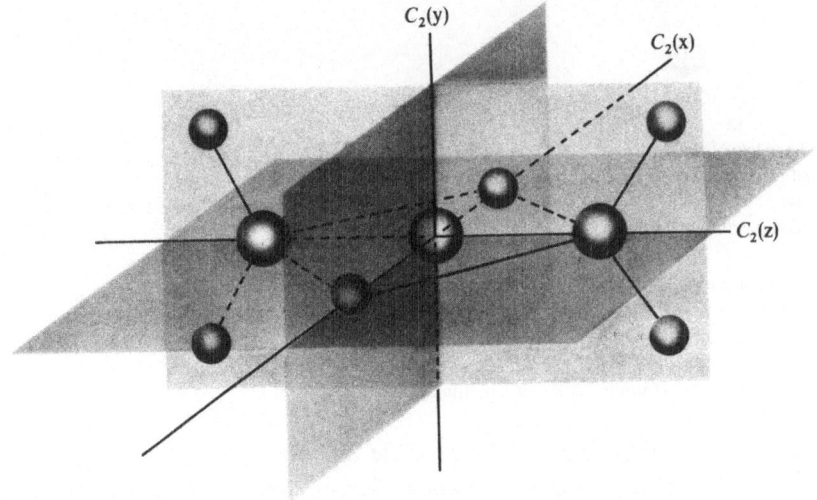

Fig. 2–1   The three two-fold axes and three planes of symmetry of diborane, which also has a centre of symmetry and belongs to the symmetry group D_{2h}

the symmetries of the bridged (D_{2h}) and ethane-like (D_{2d}) models, it should be possible to distinguish between these structures by a study of their vibrational spectra; indeed, the infrared spectrum alone can be used to discriminate between them, and it was as a result of various studies of the infrared spectrum of diborane, notably those of R. P. Bell and H. C. Longuet-Higgins,[4] and W. C. Price[5] during the nineteen-forties, that the bridged structure was confirmed. Significant features include the number of absorption bands, their frequencies, and their rotational fine structure.

The infrared spectrum of diborane (part of which is shown in Fig. 2–3) contains too many bands to be explicable in terms of an ethane-like structure, for which five infrared-active fundamental vibrations (out of a total of 12 distinct vibrational frequencies) are to be expected. The surplus bands cannot be explained away as overtones or combinations of other vibrations, being too intense and having inappropriate frequencies. However, all the bands are readily assignable in terms of the bridged D_{2h}

The three-fold proper axis and six-fold improper axis, $(C_3 : S_6)$ * and three planes of symmetry of ethane (staggered configuration), which also has a centre of symmetry and three two-fold axes of symmetry perpendicular to the three planes of symmetry, and belongs to the symmetry group D_{3d}

* An n-fold *improper* axis of symmetry (S_n) is one about which rotation through $2\pi/n$, followed by reflection in a plane at right angles to that axis, regenerates the structure

Fig. 2-2 The symmetry elements of ethane.

model, for which there are 18 possible distinct vibrations, though only eight of these involve dipole moment changes and so are infrared-active. The movements of the atoms in six of these eight fundamental vibrations are illustrated in Fig. 2–3, together with the bands in the range 900–3000 cm^{-1} to which they are assigned. The subscript numbers are those by which Bell and Longuet-Higgins identified them. Note that there are four B–H stretching modes, twice as many as would be expected for an ethane-like model. The frequencies of these absorptions, $\nu(\text{B–H}_{\text{terminal}})$ near 2600 cm^{-1} and $\nu(\text{B–H}_{\text{bridging}})$ between 1500 and 2000 cm^{-1}, are characteristic of such groups. The number and form of such bands in the spectra of other boron hydrides provides a useful indication of what type of boron-attached hydrogen is present, for example, whether bridging or terminal and whether as BH$_2$ or BH, particularly since these bands occur in regions of the spectrum relatively free from absorptions due to other groups. Moreover, the force constants that can be calculated from their frequencies show the terminal B–H links to be essentially normal single bonds, but the bridge B–H links turn out to be considerably weaker.

The assignments in Fig. 2–3 have been confirmed by studies on fully deuterated diborane, for which appropriate decreases in frequency were found. Further support is provided by the rotational fine structure of certain bands, which closely resembles that of corresponding bands in the

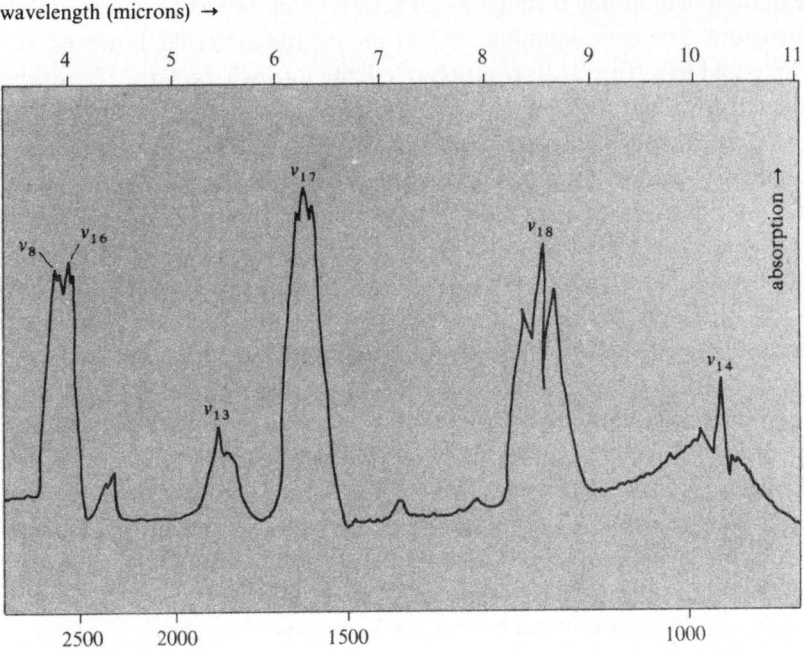

wavelength (microns) →

absorption ↑

← wave number (cm⁻¹)

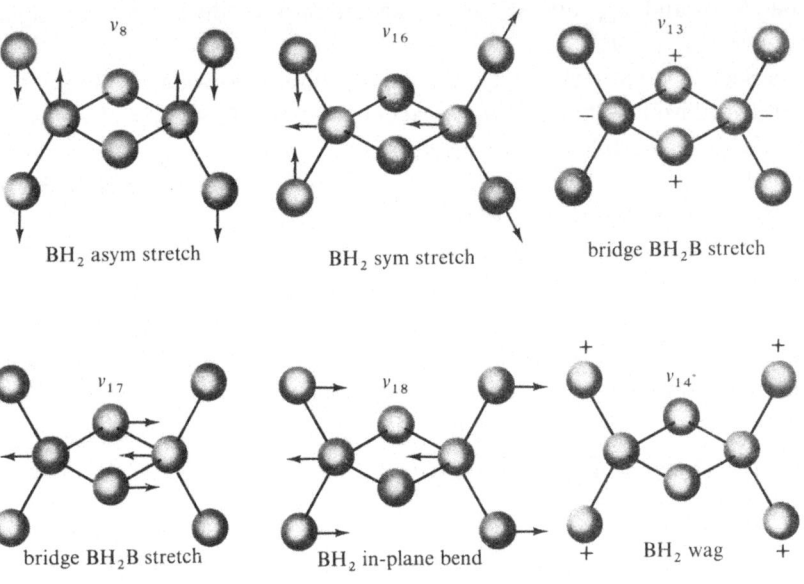

ν_8 BH₂ asym stretch

ν_{16} BH₂ sym stretch

ν_{13} bridge BH₂B stretch

ν_{17} bridge BH₂B stretch

ν_{18} BH₂ in-plane bend

ν_{14} BH₂ wag

Fig. 2–3 The infrared spectrum of diborane in the range 900–3000 cm⁻¹ and the vibrations responsible for the major absorption bands

spectrum of ethylene. Like ethylene, diborane has one two-fold axis of symmetry passing through the two heavy atoms, labelled $C_2(z)$ in Fig. 2–1. Rotation of diborane through $2\pi/2$ i.e. 180° about this axis regenerates the structure. The corresponding axis of the ethane molecule, however, is a three-fold axis (Fig. 2–2); rotation of ethane through $2\pi/3$, i.e. 120° about the axis through the two carbon atoms regenerates the structure. When the spectrum of diborane is recorded on a suitably expanded wave number scale, the rotational fine structure of the bands labelled v_8 and v_{14} in Fig. 2–3 (the asymmetrical stretching and wagging vibrations respectively of the terminal BH_2 groups) is found to have the two-fold alternation of intensity appropriate for the two-fold symmetry axis $C_2(z)$ of the D_{2h} model, but inappropriate for the three-fold symmetry axis of the ethane D_{3d} model, for which an altered intensity in every third rotational line would be expected.

With the development of infrared spectrometers capable of a resolution of 0·04–0·05 cm^{-1} at 2500 cm^{-1}, it has been possible to measure the frequencies of the rotational lines with sufficient accuracy to enable the moments of inertia of diborane to be determined extremely precisely. From these in turn can be calculated interatomic distances and angles. Such studies are aided by isotopic substitution of either boron or hydrogen. Boron occurs naturally as a mixture of the isotope ^{10}B (relative abundance 18·8%) and ^{11}B (81·2% abundance). Using samples of diborane enriched respectively in ^{10}B and ^{11}B, W. J. Lafferty, A. G. Maki, and T. D. Coyle have examined the fine structure of the two B–H (terminal) stretching bands v_8 and v_{16}, and obtained the interatomic distances and angles shown in Fig. 2–6.[6]

Such infrared studies thus quite early confirmed what had long been suspected from certain **chemical reactions** of diborane, that two hydrogens in the molecule differed from the other four. This had been shown by such reactions as those between diborane and trialkylboranes, which incidentally have the formula R_3B and are not associated. Depending on the ratio of reactants, these reactions gave a series of products, $R_nB_2H_{6-n}$, of alkyl hydrogen exchange in which n significantly never exceeded four, even when an excess of trialkylborane was used. These results are intelligible in terms of the hydrogen-bridged structure for diborane if the terminal, but not the bridging, hydrogens can be replaced by alkyl groups, for example,

$$H_2BH_2BH_2 + 4R_3B \rightarrow 3R_2BH_2BR_2$$

Further support for the bridged structure of diborane came from its **nuclear magnetic resonance spectra** first reported in 1955 (Figs 2–4 and 2–5).[7] Of the two naturally-occurring boron isotopes, the ^{11}B nucleus has spin $\frac{3}{2}$, and the ^{10}B nucleus has spin 3. Coupling involving the latter less abundant nucleus can for most purposes be ignored, as the complex signals produced are normally obscured by the more intense but simpler signals arising from coupling to ^{11}B. However, coupling between the spins of hydrogen nuclei and those of ^{11}B nuclei to which the hydrogens are directly chemically bonded provides a powerful structural probe in boron hydride chemistry. A hydrogen atom attached to a ^{11}B boron

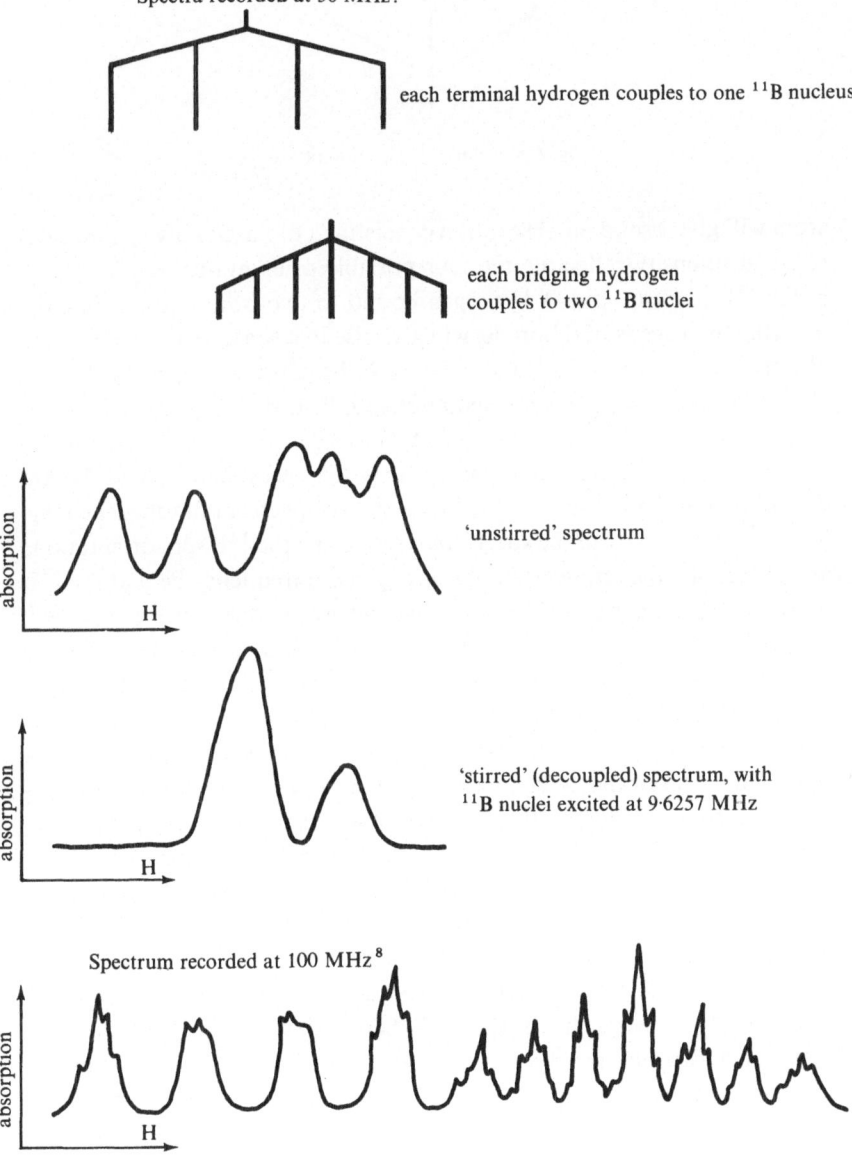

Spectra recorded at 30 MHz:

each terminal hydrogen couples to one ^{11}B nucleus

each bridging hydrogen couples to two ^{11}B nuclei

'unstirred' spectrum

'stirred' (decoupled) spectrum, with ^{11}B nuclei excited at 9·6257 MHz

Spectrum recorded at 100 MHz[8]

Fig. 2–4 ^1H n.m.r. spectra of $^{11}B_2H_6$

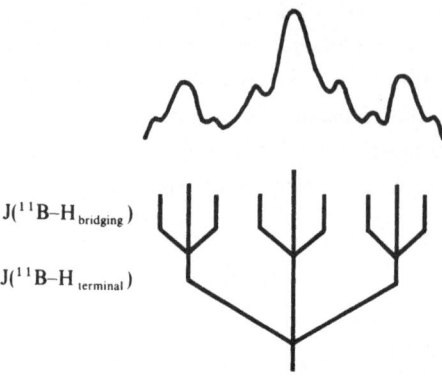

$J(^{11}B-H_{bridging})$

$J(^{11}B-H_{terminal})$

Fig. 2–5 ^{11}B n.m.r. spectrum of $^{11}B_2H_6$

atom will give rise to an absorption consisting of four equally spaced lines of equal intensity reflecting the four possible orientations $(\frac{3}{2}, \frac{1}{2}, -\frac{1}{2}, -\frac{3}{2})$ of the ^{11}B nuclear spin. A hydrogen bound to two boron atoms, like the bridging hydrogens of diborane, will give rise to a septet with components of intensities 1, 2, 3, 4, 3, 2, and 1, because the combined spins of the two ^{11}B nuclei can adopt seven orientations (3, 2, 1, 0, −1, −2, and −3 in 1, 2, 3, 4, 3, 2, and 1 ways respectively). These features, complicated by the superimposition of the two signals if they are recorded at 30 MHz, are discernable in the proton magnetic resonance spectrum of diborane (Fig. 2–4). The complexity of the spectrum is reduced if the ^{11}B spin orientations are 'stirred' by the application of a strong radio frequency field at the ^{11}B resonance frequency (9·6257 MHz) while the proton resonance is recorded, a procedure known as heteronuclear spin–spin decoupling, double irradiation, or double resonance. Under these circumstances the neighbouring protons see only the average of the ^{11}B orientations, and the fine structure due to the ^{11}B nucleus collapses to a single line. The resulting spectrum of diborane then consists of two peaks, of relative intensities 2 and 1, appropriate for the four terminal and two bridging protons. Note that the signal due to the bridging hydrogens is centred at *higher* field than that due to the terminal hydrogens, the bonding significance of which is discussed below. Note also that in the spectrum of the 'unstirred' sample, the components of the septet are much more closely spaced than are the components of the quartet, because $J(^{11}B-^1H_{terminal})$ considerably exceeds $J(^{11}B-H_{bridging})$. Separation of the bridging and terminal hydrogen signals can also be accomplished by recording the spectrum at a higher frequency. This is because the chemical shift between, i.e. separation of, resonances arising from chemically distinct atoms of a particular element in a particular molecule is an invariant fraction of the applied field (usually expressed in parts per million) whereas spin–spin coupling constants are invariant numbers of cycles per second (Hz). The higher the frequency at which a spectrum is recorded, the more compact does a multiplet absorption due to a particular atom or set of atoms appear. The high resolution spectrum recorded at 100 MHz also shown in Fig. 2–4 illustrates this.[8] The quintet fine structure of the bridging proton signals which then becomes apparent arises from their weak coupling with the four terminal protons.

In the interpretation of ^{11}B n.m.r. spectra, one simplifying feature is that ^{11}B–^{11}B spin–spin coupling effects can be ignored, as they are not detected in practice. In the ^{11}B n.m.r. spectrum of diborane (Fig. 2–5), because each boron is attached to two terminal hydrogen atoms (each of nuclear spin $\frac{1}{2}$, thus combined spins of 1, 0, and -1 of statistical weights 1, 2, and 1 respectively) and to two bridging hydrogens, and because $J(^{11}$B–^{1}H$_{terminal})$ exceeds $J(^{11}$B–^{1}H$_{bridging})$, the signal obtained consists of a triplet of triplets. The large triplet structure reflects coupling with the terminal hydrogens, the fine structure coupling with the bridging hydrogens. If diborane had had an ethane-like structure, with all six hydrogens equivalent, three attached to each boron, its ^{1}H n.m.r. spectrum would have been expected to consist of a quartet, with four equally spaced components of equal intensity, and its ^{11}B n.m.r. spectrum should also have been a quartet with the same spacings [same $J(^{11}$B–^{1}H$_{terminal})$ spin–spin coupling constant] but of relative intensities 1, 2, 2, and 1 reflecting the possible combined spins for three protons of $\frac{3}{2}, \frac{1}{2}, -\frac{1}{2}$, and $-\frac{3}{2}$ respectively.

The molecular geometry of diborane is now known with some precision, as a result of the infrared spectroscopic work already mentioned and further electron-diffraction work on the vapour.[9] The bond distances and angles derived from these two methods show excellent agreement (see Fig. 2–6). Single crystal X-ray diffraction work has also been carried

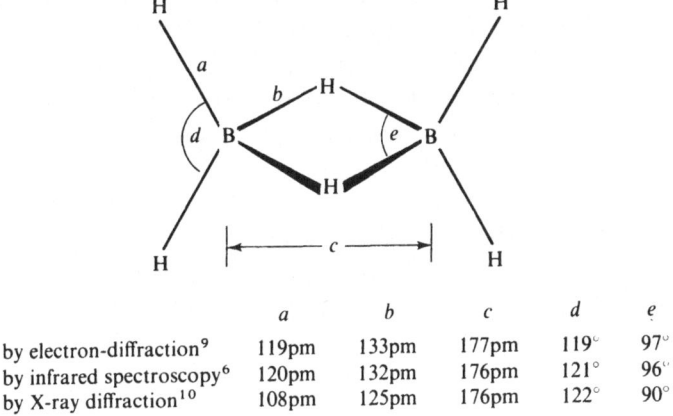

	a	b	c	d	e
by electron-diffraction[9]	119pm	133pm	177pm	119°	97°
by infrared spectroscopy[6]	120pm	132pm	176pm	121°	96°
by X-ray diffraction[10]	108pm	125pm	176pm	122°	90°

Fig. 2–6   Bond lengths and angles of diborane.

out—no mean feat on such a low melting highly-reactive substance.[10] This has given a boron–boron distance consistent with that obtained from the other studies (Fig. 2–6), but the values for the boron–*hydrogen* distances are appreciably shorter than those derived from infrared spectroscopy and electron-diffraction. This does not mean that the boron–hydrogen bonds contract when the vapour condenses. The differences arise because the different techniques are not measuring quite the same property. Whereas the interatomic distances obtained by infrared spectroscopy or electron-diffraction are essentially inter*nuclear* distances, the X-ray diffraction data reflect distances between the *centres of electronic charge*, since X-rays are scattered almost exclusively by the extranuclear electrons. For atoms of high atomic number, with a high nuclear charge and many electrons, the centre of electronic charge may be taken to coincide with the nucleus irrespectively of the state of chemical combination of those atoms. However, covalently bound hydrogen, with no other electrons than those in its valence shell, may be expected to suffer significant distortion of its electron cloud from spherical symmetry, with the centre of charge no longer coincident with the nucleus but shifted towards the other atom or atoms to which the hydrogen is bonded, whether in a terminal or bridging environment. D. S. Jones and W. N. Lipscomb[10] have calculated that some 5 pm should be added to the X-ray B–H distances listed in Fig. 2–6 to allow for such distortions from spherical symmetry, distortions which cannot be assessed without a knowledge of the distribution of bonding electrons in diborane. The development of our understanding of the bonding in diborane is outlined in the following section.

2–3   The bonding in diborane

In diborane, as Fig. 2–6 shows, each boron is at the centre of a distorted tetrahedron. The notable features of the structure, for which a bonding scheme must supply an explanation if that scheme is to be acceptable, include the following: the significant disparity between the terminal (119 pm) and bridging (133 pm) boron–hydrogen bond distances; the magnitude (122°) of the HBH bond angle between the terminally attached hydrogen atoms, and the acuteness (83°) of the BHB bond angle at the bridging hydrogen. Moreover, a satisfactory bonding scheme should also explain why diborane is stable with respect to dissociation into two borane fragments, BH_3, for which trigonal planar structures would be expected. The precise value of the enthalpy of the dissociation $B_2H_6 \rightarrow 2BH_3$ is still subject to some uncertainty, but a value in the region of 150 kJ appears likely from recent studies.

The boron–hydrogen bond lengths of diborane show where the electron deficiency lies. The terminal B–H bonds are of the same length (119 pm) as B–H bonds in other compounds for which there is no difficulty assigning a pair of electrons to each such bond. The longer bridging B–H bonds evidently involve a lower overall electron density. If a pair of electrons is assigned to each of the terminal B–H bonds, there remain altogether four valence electrons to be distributed among the four bridging B–H bonds.

One obvious way of distributing the four electrons was to allocate one to each B––H link. Indeed, four two-centre one-electron links in a BH_2B bridge were suggested by A. F. Core in 1928,[11] long before the structure of diborane was known. However, if two-centre one-electron links were to provide the key to the bonding in diborane, there were one or two awkward questions to be answered. Why were one-electron links so rare elsewhere in chemistry, being confined to such exotic species as the hydrogen molecule ion, H_2^+, discovered by Aston in his mass spectrometer? Why did they seemingly always occur in pairs in the boron hydrides, all of which are *even* electron molecules? Moreover, why were the bridging electrons spin-paired, despite their use in separate bonds? The diamagnetism of diborane implied the following spin orientations for a one-electron bonded structure:

$$
\begin{array}{ccccc}
\text{H} & & \text{H} & & \text{H} \\
\uparrow\downarrow \ \uparrow & & \downarrow & & \uparrow\downarrow \\
& \text{B} & & \text{B} & \\
\uparrow\downarrow \ \downarrow & & \uparrow & & \uparrow\downarrow \\
\text{H} & & \text{H} & & \text{H}
\end{array}
$$

Because of these difficulties, attention turned towards explanations in terms of electron-*pair* bonds, although J. W. Linnett[12] subsequently pointed out that the above description is apt in giving an immediate picture of the location and density of the bonding electronic cloud, and moreover is readily intelligible in terms of a valence description in which electrons of like spin in the valence shell of an atom are arranged as far apart as possible bearing in mind the attractive forces of the neighbouring atomic nuclei.

One bonding scheme put forward was that the two borane (BH_3) fragments of diborane were held together by an essentially electrostatic interaction between the polar $B(\delta+)$–$H(\delta-)$ bonds:

$$
\begin{array}{c}
\delta- \\
\text{H} \\
{}^{\delta+}\diagup \quad \diagdown {}^{\delta+} \\
\text{H}_2\text{B} \qquad \text{BH}_2 \\
\diagdown \quad \diagup \\
\text{H} \\
\delta-
\end{array}
$$

This explanation could be discounted on the grounds of the extremely low polarity of the B–H bonds. Boron and hydrogen differ in electronegativity by only 0·1 unit on the Pauling scale.

In principle, formal electron deficiency can always be explained in electron-pair valence-bond terms by invoking resonance between bonded and non-bonded structures or ionic forms. Still retaining the two-centre electron-pair bond approach, for example, one might describe diborane as a resonance hybrid of such covalent and ionic forms as the following:

It is not easy, however, to assess the relative contribution each of these makes, particularly as the molecular geometry, with a distorted tetrahedral arrangement of the hydrogen atoms about the boron, is inappropriate for such species as BH_3 (which would be expected to be trigonal planar) and H_2B^+ (which would be expected to be linear). Moreover, the complexity of this approach obscures the precise form of the bridge bonding.

Two of the above canonical forms, depicting diborane as $H^+B_2H_5^-$, are reminiscent of the formulation of diborane as $(H^+)_2B_2H_4^{2-}$ originally proposed in 1928 by E. Wiberg,[13] who noted the isoelectronic relationship between the anion $B_2H_4^{2-}$ and ethylene, and suggested that the two protons were in some way held within the electron shells of the boron atoms, because diborane did not behave as a dibasic acid. This formulation of diborane was resurrected by K. S. Pitzer in 1945[14] in a form that has come to be recognized as particularly apt.

Pitzer pointed out that the representation of diborane as $(H^+)_2B_2H_4^{2-}$ was consistent with all the known properties of diborane if the two protons were embedded in (and screened by) the π-electron cloud associated with the formal boron–boron double bond (Fig. 2–7). The screening of the protons would prevent their functioning as acidic hydrogens, and the length of the boron–boron bond (contrast 133 pm for $C{=}C$ in ethylene) could be ascribed to boron–boron repulsion forces arising from their formal negative charges.

The most important feature of the Pitzer π-bonded representation of

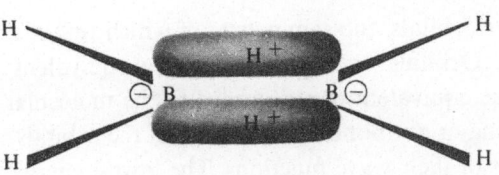

Fig. 2–7 Pitzer's protonated double bond formulation for diborane

diborane was that, in embedding the bridging hydrogen nuclei in the π-electron cloud, it associated the bridging valence electrons with both the borons *and* the bridging hydrogens, rather than locate them in pairs between particular pairs of nuclei. The logical development of this, pointed out by Longuet-Higgins (the first of whose many major contributions to this area was made while he was still an undergraduate), was to consider multicentre bonding descriptions, involving partial molecular orbitals, for the bridge region of diborane.[15]

The simplest approach is to consider the boron atoms of diborane as sp^3 hybridized. Two hybrid orbitals on each boron are used in bonding to the terminal hydrogens by essentially localized two-centre electron-pair bonds. The remaining orbitals point towards the bridging hydrogens, and interact with the latters' $1s$ orbitals. The form the interaction takes is shown in Fig. 2–8 for the three atoms constituting half the bridge of

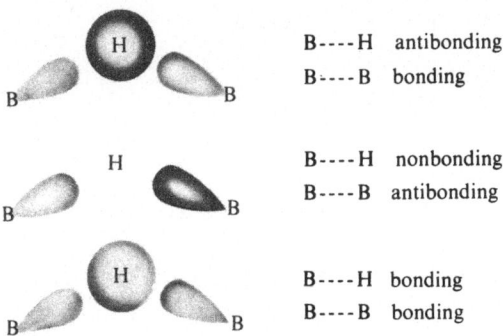

Fig. 2–8* The form of the three three-centre molecular orbitals generated from two boron sp^3 orbitals and a hydrogen $1s$ orbital

diborane, which illustrates some basic features of molecular orbital theory. For example, that a set of n atomic orbitals, interacting when their parent atoms are brought together to form a molecule, gives rise to a set

* In this and subsequent orbital diagrams, the phase of the wave function is shown thus: shaded = negative, unshaded = positive.

of *n* molecular orbitals, the symmetries of which reflect the symmetry of the molecule. Orbitals on atoms occupying equivalent positions in a molecule make equivalent contributions to the molecular orbitals.

The bonding or antibonding properties of the orbitals can be deduced from the signs of their wave functions. The lowest energy orbital formed has no change of sign of ψ between the atomic nuclei, and is therefore a bonding orbital with respect to both B———B and B———H links. The next in energy changes sign on reflection in the mirror plane through hydrogen at right-angles to the line through the boron atoms. There is zero contribution from the hydrogen 1s orbital to this orbital, since the 1s orbital is symmetric with respect to the same mirror plane. This second molecular orbital accordingly contributes nothing to B———H bonding, and is in fact *antibonding* with respect to B———B interactions. The third molecular orbital and that of highest energy, has a nodal surface in each B———H link, and so is strongly antibonding with respect to B———H interactions, though formally B———B bonding. The bonding or antibonding character of these orbitals with respect to B———H and B———B interactions is

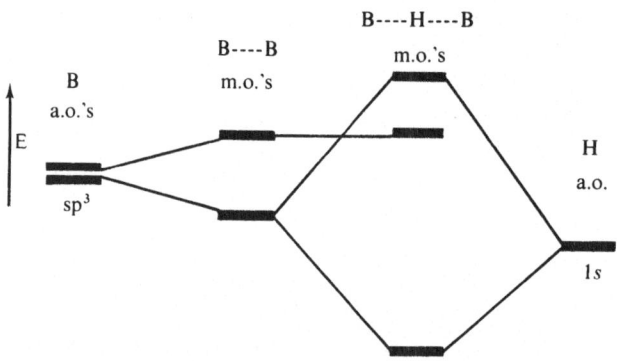

Fig. 2–9   Orbital correlation diagram for a B---H---B bridge

underlined in the correlation diagram (Fig. 2–9), which also shows how the two boron sp^3 orbitals interact weakly to give a B———B bonding and antibonding pair, only the former of which has the right symmetry to interact with the hydrogen 1s orbital. Of the three molecular orbitals, only the first is bonding, and can accommodate a pair of electrons, as can its counterpart in the other half of the BH_2B bridge. The resulting bridge bonding thus involves a total of four electrons—any more electrons would have to be accommodated in orbitals which would weaken the bridging link—accommodated in two banana-shaped orbitals which concentrate the electron cloud between the boron nuclei and around the hydrogen nuclei, as shown in Fig. 2–10. The distribution of electronic charge, with on average one electron per two-centre B———H bridging link is reminiscent

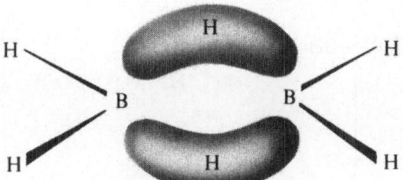

Fig. 2–10   Banana-shaped charge-clouds
of the three-centre bonds in B_2H_6

of the description of the bridge in terms of four one-electron two-centre bonds; the pairing of electronic spins required by the diamagnetism of diborane results from the occupancy of only two orbitals by the four electrons.

This description of the bonding in diborane is analogous to the description of the double bond in ethylene in terms of two bent bonds formed by overlap of carbon sp^3 hybrid orbitals, as opposed to the σ- and π-bonded treatment which follows from assumption of sp^2 hybridization. A description corresponding directly to Pitzer's protonated double bond results from assumption of sp^2 hybridization for the borons (which hybridization, as in Fig. 2–11, incidentally is better suited to the terminal

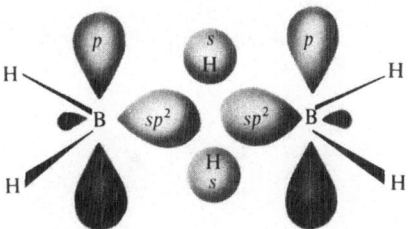

Fig. 2–11   Atomic orbitals available for
bridge bonding in B_2H_6, assuming sp^2
hybridization of the boron atoms

HBH bond angle of 122°), and treatment of the BH_2B bridge unit as a whole, instead of in two halves. This is illustrated in Fig. 2–12, which shows how the set of six bridging molecular orbitals can be built up from the superimposition of an H_2 species on a hypothetical H_2B––BH_2 species. Note that the two bonding molecular orbitals that result are both bonding with respect to boron–boron cross-ring interactions, whereas one is bonding, the other antibonding, with respect to cross-ring hydrogen–hydrogen interactions. The signs of the wave functions of the atomic orbitals that contribute to these six molecular orbitals are shown in Fig. 2–13.

It should be stressed that whether the bonding in the bridge region of diborane is described in terms of two equivalent three-centre banana-

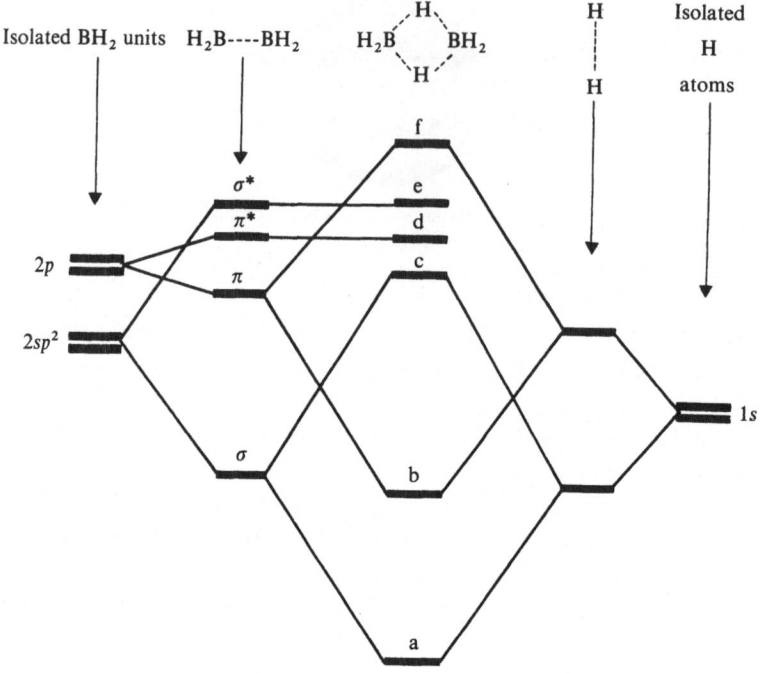

Fig. 2–12 Orbital correlation diagram for the BH_2B bridge section of diborane, assuming sp^2 hybridization

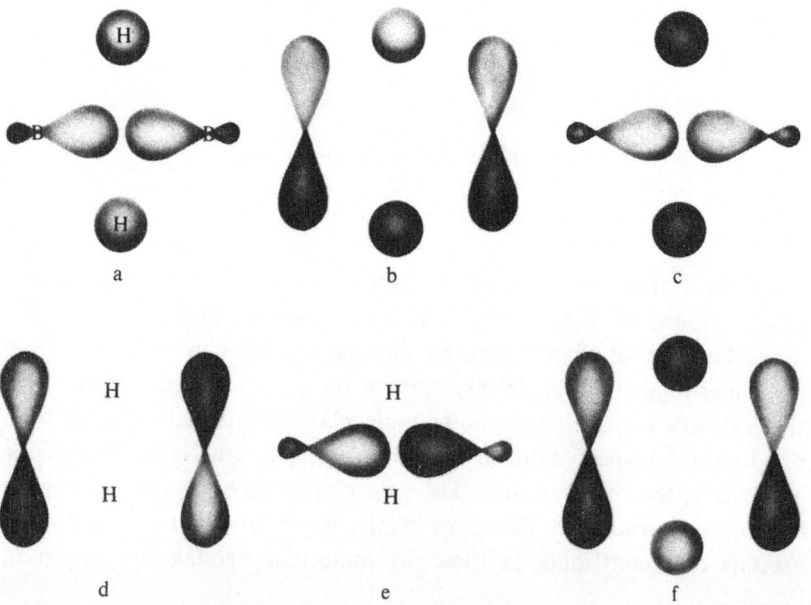

Fig. 2–13 The combinations of atomic orbitals that generate the six four-centre bridge orbitals of Fig. 2–12

shaped bonds, or in terms of two four-centre orbitals (those labelled a and b in Figs 2–12 and 2–13), the overall distribution of electronic charge in the bonding region is the same in both cases, concentrated mainly around the hydrogen nuclei and between them and the boron nuclei rather than on the B–B axis. The average of one electron per bridging B---H bond is consistent with the length of these bonds; the greater degree of boron–boron than of hydrogen–hydrogen bonding across the ring accounts for the acute angle at hydrogen; the electrons are paired, as required by the diamagnetism of diborane. Moreover, the molecule has just enough electrons to fill all the bonding orbitals—any extra electrons would have to be accommodated in a weakly antibonding orbital. Given an extra pair of electrons, in fact, the molecule would be expected to rearrange to the ethane-like structure, which is that appropriate for 14 bonding electrons. The stability of diborane with respect to dissociation into two borane fragments does not follow *automatically* from this model, but requires a knowledge of the precise energy levels of the bonding molecular orbitals in both diborane and borane.

Although sp^2 hybridization was assumed for the boron atoms in constructing the set of molecular orbitals shown in Figs 2–12 and 2–13, a similar set of molecular orbitals, only two of which are bonding (Fig. 2–14), results if one assumes sp^3 hybridization. This description therefore is not an *explanation* of the value of the HBH bond angle between the terminal

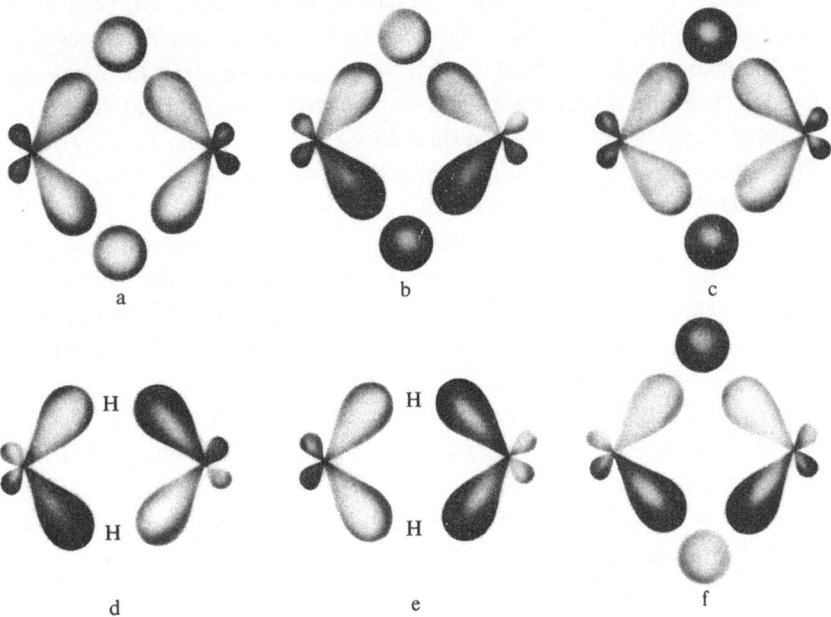

Fig. 2–14 The combinations of boron $2sp^3$ and hydrogen $1s$ atomic orbitals that generate a set of six four-centre bridge orbitals analogous to those shown in Fig. 2–13

B–H bonds. That this angle is greater than the tetrahedral angle is consistent with the valence shell electron-pair repulsion approach to bond angles, in which it is argued that bond angles reflect the interelectronic repulsion forces within the valence shells of atoms. The greater electron density near boron in the terminal B–H bonds than in the bridging B–H bonds would cause the terminal HBH bond angle to expand, and the bridging HBH bond angle to contract, relative to undistorted tetrahedral geometry. Though the actual geometry of diborane is intelligible on this basis, this does not mean that interelectronic repulsion forces alone are responsible for the observed bond angles, which (as in all molecules) represent a balance between the various internuclear and interelectronic repulsion forces and nuclear-electronic attractive forces.

The three-centre and four-centre molecular orbital treatments of the bridging region of diborane outlined above are *partial* molecular orbital treatments devised to give an indication of the number and distribution of electrons in the bridging region. Note that whether a three- or a four-centre treatment is used, and whether sp^2 or sp^3 hybridization of the boron atoms is assumed, the interaction between the available atomic orbitals *that is dictated by the symmetry of the bridge* determines how many molecular orbitals will result that are lower in energy than the parent atomic orbitals, i.e. that are *bonding* orbitals. The precise energy sequence of the higher energy (antibonding or essentially non-bonding) molecular orbitals shown in Figs 2–9 and 2–12 cannot be deduced solely from this qualitative approach, as this will depend on the interatomic distances and on the energy gaps between the hydrogen $1s$, boron $2s$, and boron $2p$ energy levels. For an understanding of the distribution and relative energy levels of *all* the bonding electrons in diborane, it is necessary to carry out a full molecular orbital treatment in which the symmetry of the whole molecule (D_{2h}) is taken into account.

Firstly, let us consider what atomic orbitals are available. No preliminary hybridization of the boron atomic orbitals is assumed. There are altogether 14 atomic orbitals from which therefore 14 molecular orbitals may be constructed. Each hydrogen atom contributes one $1s$ orbital, which for convenience will be labelled h_1, h_2, and so on, the subscript identifying the hydrogen atom according to the numbering system shown in Fig. 2–15. Each boron atom contributes one $2s$ orbital, denoted s, and three p orbitals denoted x, y, and z according to their spatial orientation relative to the axes (see Fig. 2–15), again with subscripts to indicate at which boron they are centred.

Secondly, let us consider in what way these atomic orbitals can interact. The combinations in which they can be used to construct molecular orbitals must reflect the symmetry of diborane. For example, because the two boron atoms occupy equivalent positions in the molecule, if s_1

Orbitals in the yz plane, including the terminal hydrogens

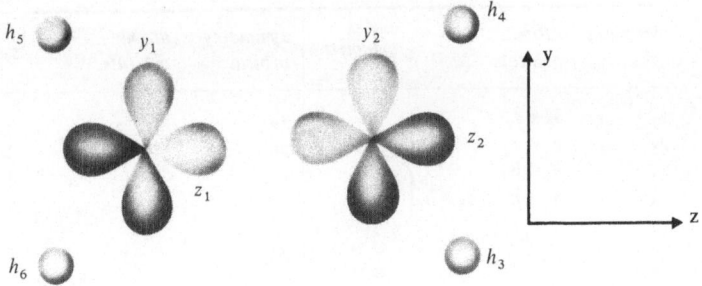

Orbitals in the xz plane, including the bridging hydrogens

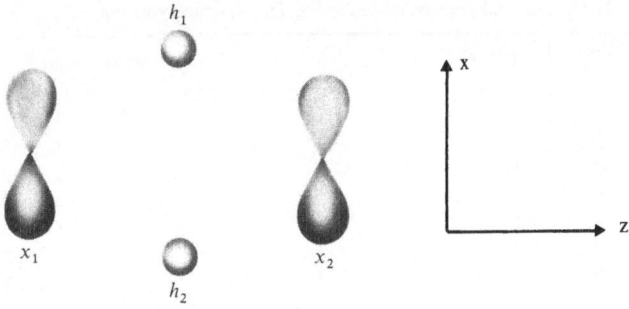

(the boron 2s orbitals are not shown in either diagram)

Fig. 2–15 The valence shell atomic orbitals from which the molecular orbitals of diborane may be constructed

contributes to a molecular orbital, so must s_2; the two combinations $s_1 + s_2$ and $s_1 - s_2$ are allowed. Similar arguments permit the combinations $x_1 + x_2$, $x_1 - x_2$, $y_1 + y_2$, $y_1 - y_2$, $z_1 + z_2$, $z_1 - z_2$, $h_1 + h_2$, and $h_1 - h_2$. Because there are four terminal hydrogen atoms, contributions from their 1s orbitals can take the four forms $h_3 + h_4 + h_5 + h_6$, $h_3 + h_4 - h_5 - h_6$, $h_3 - h_4 + h_5 - h_6$, and $h_3 - h_4 - h_5 + h_6$. Combinations such as $h_3 + h_4 + h_5 - h_6$ are excluded, because they single out one hydrogen as distinct from the remaining three, for which there is no symmetry justification. In this way the atomic orbitals can be grouped into sets, $s_1 + s_2$, $s_1 - s_2$, $x_1 + x_2$, and so on, known as **symmetry orbitals,** which simplify the problem of deciding what interactions between different types of atomic orbital may occur to generate molecular orbitals. These 14 symmetry orbitals, given the labels $\chi_1, \chi_2, \cdots \chi_{14}$ for purposes of identification, are listed in Table 2–1.

The third step is to examine in what way these symmetry orbitals may interact with each other. Interactions can occur only between symmetry

Table 2-1 Symmetry classification of the atomic orbitals of diborane

symmetry orbital	atomic orbitals	symmetry	symmetry orbital	atomic orbitals	symmetry
χ_1	s_1+s_2		χ_8	s_1-s_2	
χ_2	z_1+z_2	a_g	χ_9	z_1-z_2	b_{1u}
χ_3	h_1+h_2		χ_{10}	$h_3+h_4-h_5-h_6$	
χ_4	$h_3+h_4+h_5+h_6$				
χ_5	x_1-x_2	b_{2g}	χ_{11}	y_1+y_2	b_{2u}
			χ_{12}	$h_3-h_4-h_5+h_6$	
χ_6	y_1-y_2	b_{3g}	χ_{13}	x_1+x_2	b_{3u}
χ_7	$h_3-h_4+h_5-h_6$		χ_{14}	h_1-h_2	

Table 2-2 Character table for the D_{2h} symmetry group

D_{2h}	E	$C_2(z)$	$C_2(y)$	$C_2(x)$	i	$\sigma(xy)$	$\sigma(xz)$	$\sigma(yz)$		
A_g	1	1	1	1	1	1	1	1		x^2, y^2, z^2
B_{1g}	1	1	-1	-1	1	1	-1	-1	R_z	xy
B_{2g}	1	-1	1	-1	1	-1	1	-1	R_y	xz
B_{3g}	1	-1	-1	1	1	-1	-1	1	R_x	yz
A_u	1	1	1	1	-1	-1	-1	-1		
B_{1u}	1	1	-1	-1	-1	-1	1	1	z	
B_{2u}	1	-1	1	-1	-1	1	-1	1	y	
B_{3u}	1	-1	-1	1	-1	1	1	-1	x	

Table 2-3 The 14 valence shell molecular orbitals of diborane

Orbital	Symmetry label*
$\phi_1 = c_{11}\chi_1 - [c_{12}\chi_2] + c_{13}\chi_3 + c_{14}\chi_4$	$2a_g$
$\phi_2 = [c_{21}\chi_1] + c_{22}\chi_2 + c_{23}\chi_3 - c_{24}\chi_4$	$3a_g$
$\phi_3 = [c_{31}\chi_1] - c_{32}\chi_2 - c_{33}\chi_3 + c_{34}\chi_4$	$4a_g$
$\phi_4 = c_{41}\chi_1 + [c_{42}\chi_2] - c_{43}\chi_3 - c_{44}\chi_4$	$5a_g$
$\phi_5 = c_{55}\chi_5$	$1b_{2g}$
$\phi_6 = c_{66}\chi_6 + c_{67}\chi_7$	$1b_{3g}$
$\phi_7 = c_{76}\chi_6 - c_{77}\chi_7$	$2b_{3g}$
$\phi_8 = c_{88}\chi_8 + c_{89}\chi_9 + c_{810}\chi_{10}$	$2b_{1u}$
$\phi_9 = c_{98}\chi_8 - c_{99}\chi_9 - c_{910}\chi_{10}$	$3b_{1u}$
$\phi_{10} = [c_{108}\chi_8] - c_{109}\chi_9 + c_{1010}\chi_{10}$	$4b_{1u}$
$\phi_{11} = c_{1111}\chi_{11} + c_{1112}\chi_{12}$	$1b_{2u}$
$\phi_{12} = c_{1211}\chi_{11} - c_{1212}\chi_{12}$	$2b_{2u}$
$\phi_{13} = c_{1313}\chi_{13} + c_{1314}\chi_{14}$	$1b_{3u}$
$\phi_{14} = c_{1413}\chi_{13} - c_{1314}\chi_{14}$	$2b_{3u}$

* The designations $1a_g$ and $1b_{1u}$ do not appear in this table, being reserved for the two orbitals formed by interactions involving the boron 1s orbitals.
Terms in parentheses make only small contributions to the molecular orbital.

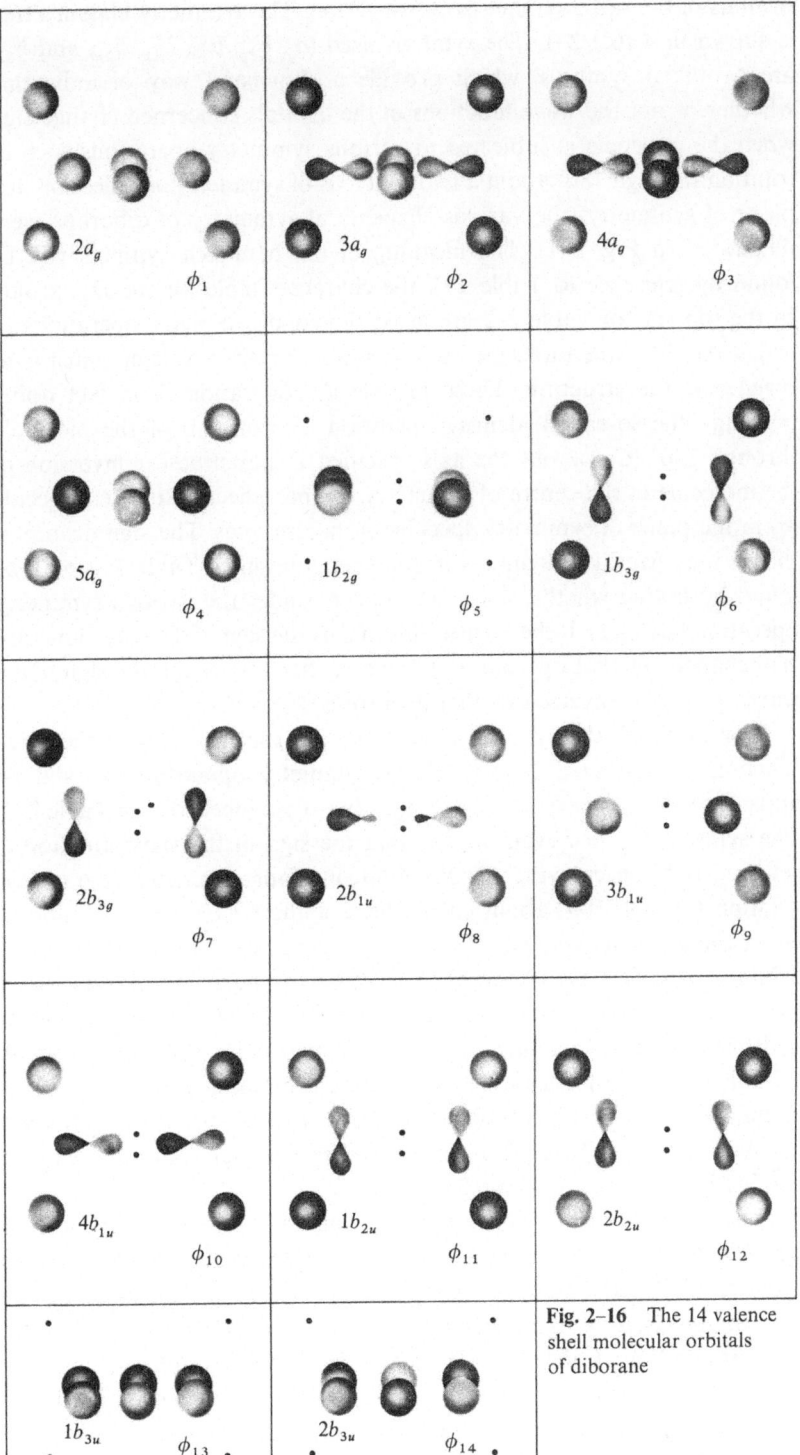

Fig. 2–16 The 14 valence shell molecular orbitals of diborane

orbitals of the same symmetry classification. The symmetry classification is shown in Table 2–1. The symbols used (a_g, b_{2g}, b_{3g}, b_{1u}, b_{2u}, and b_{3u}) are Mulliken symbols, which provide a shorthand way of indicating whether or not the wave functions of the orbitals concerned change sign when the molecule is subjected to various symmetry operations such as rotation through 180° about a two-fold axis of symmetry or reflection in a plane of symmetry (the various elements of symmetry of diborane were illustrated in Fig. 2–1). The meaning of the Mulliken symbols will be found by reference to Table 2–2, the character table for the D_{2h} group. In the top row of Table 2–2 are listed the eight symmetry operations to which the diborane molecule may be subjected, the result of which is to regenerate the structure. These include the operation of in fact doing nothing—the so-called identity operation, E; rotation of the molecule through 180° (C_2) about the axis specified in parentheses; inversion of the molecule in the centre of symmetry, i; and reflection of the molecule (σ) in the plane of symmetry specified in parentheses. The significance of the various Mulliken symbols in the first column of Table 2–2 may be found by noting whether 1 or -1 appears under the various symmetry operation headings. If the former, this means the sign of the wave function is unchanged by that operation; if however the entry is -1, the sign of the wave function is reversed by that operation.

For example, the symbol A_g or a_g means that the sign of the wave function is unaffected by any of the symmetry operations, as will be apparent for the four symmetry orbitals ($\chi_1 \rightarrow \chi_4$) labelled a_g in Table 2–1. The symbol B_{2g}, however, implies that the sign of the wave function is reversed only on carrying out the following four symmetry operations; rotation through 180° about either the z or the x axis, and reflection in either the xy or the yz plane. This will be seen to be true of the symmetry orbital $x_1 - x_2$ (χ_5), the π^* combination of the two boron p_x orbitals, which incidentally has a symmetry which no combination of other boron and/or hydrogen orbitals can have, and so this particular symmetry orbital, suitably normalized, constitutes one of the antibonding molecular orbitals of diborane (ϕ_5 in Fig. 2–16). The correctness of the other symmetry symbols given in Table 2–1 for the other symmetry orbitals can be verified by reference to Table 2–2.

The fourth step is to form combinations of those symmetry orbitals which have the same symmetry. As an example, let us consider the two symmetry orbitals that have symmetry b_{3u}. These are χ_{13} ($x_1 + x_2$, the π combination of boron p orbitals in the bridge plane) and χ_{14} ($h_1 - h_2$, the antibonding, bridge H$\cdots$H cross-ring combination). Adding and subtracting these in turn, we arrive at the two b_{3u} molecular orbitals (ϕ_{13} and ϕ_{14}) shown in Fig. 2–16. These are the bonding combination $c_{1313}\chi_{13} + c_{1314}\chi_{14}$ (where c_{1313} and c_{1314} are constants which determine the relative contributions of χ_{13} and χ_{14}; this orbital is the same as orbital b of Fig. 2–13); and the antibonding combination χ_{14} ($c_{1413}\chi_{13} - c_{1414}\chi_{14}$; orbital f of Fig. 2–13). Applying the same argument to the two symmetry orbitals of symmetry b_{3g} we obtain the terminal B–H bonding and antibonding molecular orbitals (ϕ_6 and ϕ_7) in Table 2–3 and Fig. 2–16. The terminal B–H bonding and antibonding molecular orbitals of symmetry b_{2u} (χ_{11} and χ_{12}), also shown in Fig. 2–16, result from the combinations $c_{1111}\chi_{11} + c_{1112}\chi_{12}$ and $c_{1211}\chi_{11} - c_{1212}\chi_{12}$.

The remaining molecular orbitals shown in Fig. 2–16 and listed in Table 2–3 are constructed from the four symmetry orbitals (χ_1, χ_2, χ_3, and χ_4) of symmetry a_g, and the three (χ_8, χ_9, and χ_{10}) of symmetry b_{1u}. From these, two bonding molecular orbitals of a_g symmetry, and one of b_{1u} symmetry, can be generated. The remaining four orbitals are antibonding orbitals. Considering firstly the bonding molecular orbitals of a_g symmetry, the most strongly bonding combination of these (ϕ_1) may be represented by $c_{11}\chi_1 - c_{12}\chi_2 + c_{13}\chi_3 + c_{14}\chi_4$, where again the constants c_{11} and so on, determine the relative contribution made by each symmetry orbital χ. Because the boron $2s$ atomic orbitals s_1 and s_2 are lower in energy than the boron $2p_z$ atomic orbitals z_1 and z_2, the latter contribute much less than the former ($c_{12} \ll c_{11}$). The effect of the small contribution from the p_z orbitals is to introduce some sp hybrid character into the contribution made by the boron orbitals, and so to increase slightly the bonding of the boron atoms to the terminal hydrogen atoms, at the expense of bonding to the bridge hydrogens. Conversely, the other bonding molecular orbital of a_g symmetry, ϕ_2, which may be represented by $c_{21}\chi_1 + c_{22}\chi_2 + c_{23}\chi_3 - c_{24}\chi_4$, has but a slight contribution from the boron $2s$ orbitals ($c_{21} \ll c_{22}$), and is essentially as shown ($3a_g$) in Fig. 2–16. The one bonding molecular orbital of b_{1u} symmetry ($\phi_8 = c_{88}\chi_8 + c_{89}\chi_9 + c_{810}\chi_{10}$) has comparable contributions from χ_9 and χ_{10}, so may be regarded as a terminal B–H bonding orbital constructed from the terminal hydrogen $1s$ orbitals and sp hybrid orbitals on the borons. These and other features of the 14 molecular orbitals of diborane are illustrated in Fig. 2–16. Note that, of the six bonding orbitals (ϕ_1, ϕ_2, ϕ_6, ϕ_8, ϕ_{11}, and ϕ_{13}), those of symmetry a_g (ϕ_1 and ϕ_2) contribute to the bonding of both terminal and bridging hydrogens, those of symmetry b_{3g} (ϕ_6), b_{1u} (ϕ_8), and b_{2u} (ϕ_{11}) bond exclusively terminal hydrogens, and that of symmetry b_{3u} (ϕ_{13}) bonds exclusively the bridge hydrogens.

To summarize: arguing solely from symmetry, we conclude that the symmetry-dictated interactions between the 14 valence-shell orbitals of the eight constituent atoms of diborane will lead to six bonding molecular orbitals, two of a_g symmetry, and one each of b_{3g}, b_{1u}, b_{2u}, and b_{3u}. The energy sequence of these molecular orbitals cannot be deduced from this approach, as this is not symmetry-dictated, but depends on the interatomic distances. Using these distances, various nonempirical calculations of the molecular orbital energy levels have been carried out by determining what values of the various parameters c_{11} and so on, i.e. what precise **linear combinations** of **atomic orbitals,** generate the field for which they are appropriate, i.e. give a **self-consistent field** (the so-called SCF–LCAO approach). An approximate set of orbitals is considered, from which the electrostatic field acting on each electron is calculated. This field is then used to calculate new orbitals from which improved approximations to the field are obtained, and so on until the set of orbitals regenerates the field that gave these orbitals. From such calculations, it emerges[16] that for the bonding orbitals the sequence is

$$1a_g(\phi_1) < 1b_{1u}(\phi_8) < 1b_{3u}(\phi_{13}) < 1b_{2u}(\phi_{11}) < 2a_g(\phi_2) < 1b_{3g}(\phi_6)$$

Probably the most interesting feature of this sequence is that the b_{3u} orbital ϕ_{13}, which corresponds to the protonated π-bond of Pitzer's description, is lower in energy than three of the σ-bonding orbitals, whereas this π-bonding molecular orbital is the highest energy filled orbital in ethylene. Removal of an electron from the highest occupied orbital of diborane may be expected to weaken the *terminal* B–H bonds. This sequence of orbitals has been confirmed by a study of the **photoelectron spectrum** of diborane, by which the ionization energy needed to remove an electron from each of these levels has been measured.[17] Irradiation of diborane with photons of high energy (hv), such as those of energy 40·8 eV (3940 kJ) generated in the vacuum ultraviolet region of the spectrum by a helium discharge lamp, causes electrons to be emitted from the valence shell of the diborane. The energy of these emitted electrons (E) represents the amount by which the energy of the irradiating photons (hv) exceeds the ionization energy of the electrons (I). Measurement of E accordingly allows calculation of I by the simple relationship

$$I = hv - E$$

The photoelectron spectra of diborane and ethylene are shown in Fig. 2–17, where the energy levels of the molecular orbitals of these isoelectronic molecules are correlated. Pitzer's description of diborane, emphasizing the relationship between the hypothetical anion $B_2H_4^{2-}$ and ethylene, is seen to be most apt. The molecular orbitals of unprotonated $B_2H_4^{2-}$ would be expected to be analogous in symmetry and energy sequence to those of ethylene, though of higher energy because of the lower nuclear charge of boron than of carbon. Embedding the bridging protons in the double bond to form diborane preferentially stabilizes the π-molecular orbital. It incidentally also places these two hydrogen nuclei in a region of electron density very similar to that surrounding the terminal hydrogens. Indeed, the fact that the absorption due to the bridging protons occurs at *higher*

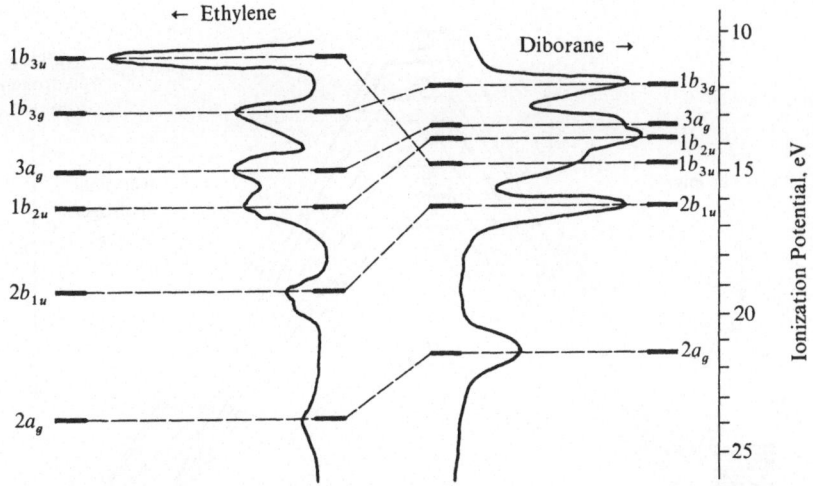

Fig. 2–17 The photoelectron spectra of diborane and ethylene (after Brundle, C. R., M. B. Robin. H. Basch. M. Pinsky. and A. Bond. ref. 17)

field than that due to the terminal protons in the ^1H nuclear magnetic resonance spectrum of diborane was originally taken as evidence that the bridging protons were in a region of greater electron density than the terminal protons, i.e. the bridging hydrogens were more hydridic ($H^{\delta-}$) in character than the terminal hydrogens, and the results of some early calculations tended to support this view. More recent calculations indicate that the electron density at the bridging hydrogens may be slightly greater than at the terminal hydrogens, although the bridging hydrogens may nevertheless have a slight overall positive charge, and the terminal hydrogens a slight overall negative charge.

One further probe of electronic energy levels which is not particularly informative in the case of diborane is the **ultraviolet absorption spectrum**,[5] in which three peaks have been observed in the region before an absorption continuum is reached. That of lowest energy, at 6·8 eV (660 kJ) with an absorption coefficient of 25, probably arises from a forbidden transition; the other two are much more intense ($\varepsilon \sim 10000$) and are located at 9·2 and 10·3 eV (890 and 990 kJ). Suggested assignments for these bands are $1b_{3g}(\phi_6) \rightarrow 1b_{2g}(\phi_5)$, i.e. a $\sigma \rightarrow \pi^*$ transition, for the lowest energy band; $1b_{3u}(\phi_{13}) \rightarrow 1b_{2g}(\phi_5)$, i.e. $\pi \rightarrow \pi^*$, for the next band; and $3a_g(\phi_2) \rightarrow 3b_{1u}(\phi_9)$ ($\sigma \rightarrow \sigma^*$) for the third band.

The electronic spectrum thus tells us little about the relative energies of the virtual orbitals of diborane—the unoccupied antibonding orbitals ϕ_3, ϕ_4, ϕ_5, ϕ_7, ϕ_9, ϕ_{10}, ϕ_{12}, and ϕ_{14} of Table 2–3 and Fig. 2–16. However, calculations of their energies have been made by various workers. The lowest unoccupied orbital is ϕ_5, the π^* combination of the boron $2p_x$ orbitals. The sequence illustrated in the orbital correlation diagram

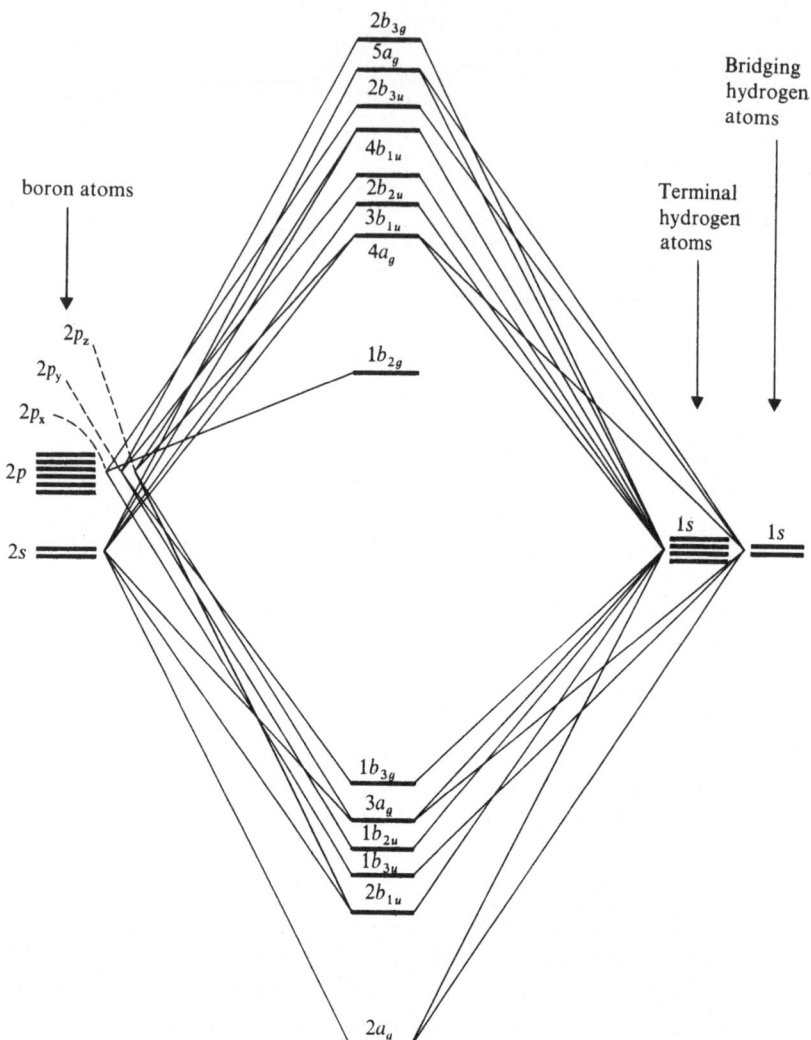

Fig. 2–18   Valence shell orbital correlation diagram for diborane

shown in Fig. 2–18 is that resulting from calculations by E. Switkes, R. M. Stevens, W. N. Lipscomb, and M. D. Newton.[16] Although minor positional adjustments of the antibonding orbitals may result from further work the diagram illustrates the large energy gap between the highest filled (ϕ_6) and lowest unfilled (ϕ_5) orbitals, which emphasizes the 'closed shell' configuration of the diborane molecule.

References

1. Stock, A., and C. Massenez, *Ber.*, 1912, **45**, 3529; Stock, A., and K. Friederici, *Ber.*, 1913, **46**, 1959; Stock, A., K. Friederici, and O. Priess, *ibid.*, p. 3353.
2. Dilthey, W., *Z. angew. Chem*, 1921, **34**, 596.
3. Mark, H., and E. Pohland, *Z. Krist.*, 1925, **62**, 103.

4. Bell, R. P., and H. C. Longuet-Higgins, *Proc. Roy. Soc.*, 1945, **183**, 357.
5. Price, W. C., *J. Chem. Phys.*, 1948, **16**, 894.
6. Lafferty, W. J., A. G. Maki, and T. D. Coyle, *J. Mol. Spectroscopy*, 1970, **33**, 345.
7. Shoolery, J. N., *Discussions Faraday Soc.*, 1955, **19**, 215.
8. Farrar, T. C., R. B. Johannesen, and T. D. Coyle, *J. Chem. Phys.*, 1968, **49**, 281.
9. Bartell, L. S., and B. L. Carroll, *J. Chem. Phys.*, 1965, **42**, 1135; see also Kuchitsu, K., *ibid.*, 1968, **49**, 4456.
10. Smith, H. W., and W. N. Lipscomb, *J. Chem. Phys.*, 1965, **43**, 1060; Jones, D. S., and W. N. Lipscomb, *ibid.*, 1969, **51**, 3133.
11. Core, A. F., *J. Soc. Chem. Ind.*, 1927, **46**, 642; *Chem. Ind. Rev.*, 1927, **5**, 642.
12. Linnett, J. W., *The Electronic Structure of Molecules*, Methuen, London, 1964, p. 120.
13. Wiberg, E., *Z. anorg. allgem. Chem.*, 1928, **173**, 199.
14. Pitzer, K. S., *J. Amer. Chem. Soc.*, 1945, **67**, 1126.
15. Longuet-Higgins, H. C., *J. Chim. phys.*, 1949, **46**, 275.
16. Yamazaki, M., *J. Chem. Phys.*, 1957, **27**, 1401; Burnelle, L., and J. J. Kaufman, *ibid.*, 1965, **43**, 3540; Buenker, R. J., S. D. Peyerimhoff, L. C. Allen, and J. L. Whitten, *ibid.*, 1966, **45**, 2835; Switkes, E., R. M. Stevens, W. N. Lipscomb, and M. D. Newton, *ibid.*, 1969, **51**, 2085.
17. Brundle, C. R., M. B. Robin, H. Basch, M. Pinsky, and A. Bond, *J. Amer. Chem. Soc.*, 1970, **92**, 3863.

General references

Longuet-Higgins, H. C., *Quart. Rev.*, 1957, **11**, 121.

Problems

2-1 Verify the symmetry classifications of the sets of atomic orbitals in Table 2–1 by referring to Table 2–2.

2-2 Classify the vibrations shown in Fig. 2–3 according to their symmetries, by examining whether the direction of movement of the atoms is unchanged or reversed on carrying out the symmetry operations listed in Table 2–2.

2-3 Construct an orbital correlation diagram, of the type shown in Fig. 2–12, to show how the molecular orbitals depicted in Fig. 2–14 arise from interactions between two BH_2 species in which the boron atoms are treated as sp^3 hybridized, and two hydrogen atoms.

2-4 Discuss the possibility that diborane might have had a structure with two terminal BH groups and four bridging hydrogens located at the corners of a square at right angles to the HB————BH axis.

2-5 Describe and explain the main features that would be expected in the ^1H and ^{11}B n.m.r. spectra of the following species (assuming that suitable samples could be prepared):
  (a) Borane, BH_3.
  (b) The diborane-(6) anion, $B_2H_6^{2-}$.
  (c) A compound containing two BH_3 groups linked by a single bridging hydrogen, as in the anion $H_3BHBH_3^-$.
  (d) A compound containing two BH_2 groups linked by a single bridging hydrogen.

2-6 Explain how a consideration of the symmetry-dictated interactions between the valence-shell atomic orbitals of the constituent atoms of diborane can be used to deduce the number and symmetries of its bonding molecular orbitals.

  [Use the character table for the D_{2h} group (Table 2–2). In order to determine in what manner the choice of axes influences the result, take the z axis of the molecule as that passing through the boron atoms, and the y axis as that passing through the bridging hydrogen atoms. Compare your conclusions with those in Tables 2–1 and 2–3.]

3 Structures and bonding of some higher boranes and boron polyhedra

3–1 Introduction

We saw in Chapter 2 how the bridge bonding in diborane could be described simply in terms of two banana-shaped three-centre electron-pair bonds, although an understanding of the overall distribution of the valence electrons of diborane required a molecular orbital treatment which took into account the symmetry of the whole molecule. The concept of three-centre electron-pair bonds can profitably be applied to a wide variety of other electron deficient compounds. W. N. Lipscomb,[1] for example, has shown how its use allows a relatively simple bonding interpretation to be made of the seemingly complicated structures of most of the higher hydrides of boron, as outlined below. Not all electron deficient compounds of boron have structures that are readily interpreted in terms of simple three-centre bonded schemes, however, and the second half of this chapter is concerned with some cage structures that figure in the hydride, halide, and alloy chemistry of boron as well as in the element itself, structures in which boron atoms are located at the corners of the equilateral triangular faces of polyhedra containing no bridging hydrogens. Bonding schemes for such structures are best arrived at by more or less full molecular orbital treatments.

3–2 The structures and bonding of some higher boranes

Boron forms a large number of higher hydrides, of general formula B_pH_{p+q}. These are named[2] in the same way as diborane, with a prefix indicating the number of boron atoms in the molecule and where necessary a numerical suffix indicating the number of hydrogens. For example, the compound B_4H_{10} is known as tetraborane(10) or simply as tetraborane, there being no other known stable binary hydride of boron containing four boron atoms. Two pentaboranes are known, however, and so are referred to as pentaborane(9) (B_5H_9) and pentaborane(11) (B_5H_{11}). The range of boron hydrides that have been isolated and characterized continues to grow, and it is possible here to survey only a selection, chosen to illustrate the structural features that are typical of these compounds and the bonding schemes in terms of which these structural features can be understood.

The description of the bridge bonding in diborane in terms of two BHB units, each held together by a three-centre electron-pair bond, is particularly useful because the structural feature in which a single hydrogen atom acts as a bridge between two boron atoms is common to all the higher boron hydrides whose structures have been determined, and the bonding description used for diborane can conveniently be applied to such cases. Another feature shared by many higher boranes is one in which another boron atom, rather than a hydrogen atom, functions as the bridging species between two borons. Two types of **BBB** three-centre bonding may be envisaged, depending on whether the bridging boron is regarded as contributing an sp^3 hybrid atomic orbital pointing towards the centre of the triangle of boron atoms, as in Fig. 3–1, or a p orbital at right

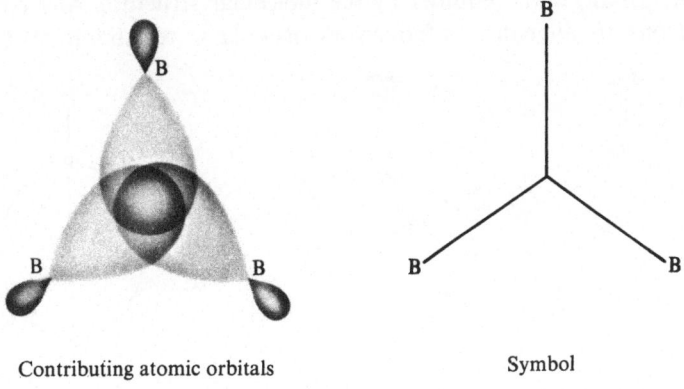

Contributing atomic orbitals   Symbol

Fig. 3–1   Closed three-centre BBB bond

angles to this, as in Fig. 3–2. The former is referred to as a 'closed', and the latter as an 'open', three-centre bond. In some compounds, the arrangement of the other atoms about the central bridging boron may be taken as a guide to which of the two three-centre **BBB** bonding representations is more appropriate. In others, either the open or the closed three-

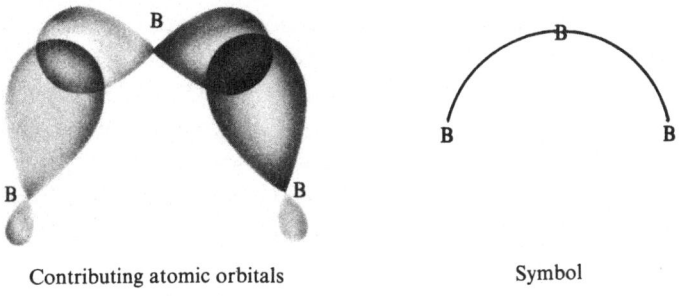

Contributing atomic orbitals   Symbol

Fig. 3–2   Open three-centre BBB bond

centre BBB bond may be compatible with the molecular geometry, and resonance between various canonical forms may be invoked. The use of such three-centre bonds has enabled bonding schemes to be devised for all known and some hypothetical boron hydrides, and is illustrated here for the compounds B_4H_{10}, B_5H_{11}, B_6H_{10}, and $B_{10}H_{14}$.

Their structures, all of which have been determined by X-ray crystallography, are illustrated in Figs 3–3, 3–5, 3–8, and 3–10.

The bonding schemes also shown have been arrived at assuming that each hydrogen contributes a $1s$ orbital, and each boron a $2s$ and three $2p$ orbitals, hybridized as appears appropriate for its coordination. Electrons (one from each hydrogen atom, three from each boron) are allocated in pairs to each of the terminal two-centre B–H bonds and bridging three-centre BHB bonds required by the molecular structure. Any remaining electrons are allocated to framework orbitals. In tetraborane(10), B_4H_{10}

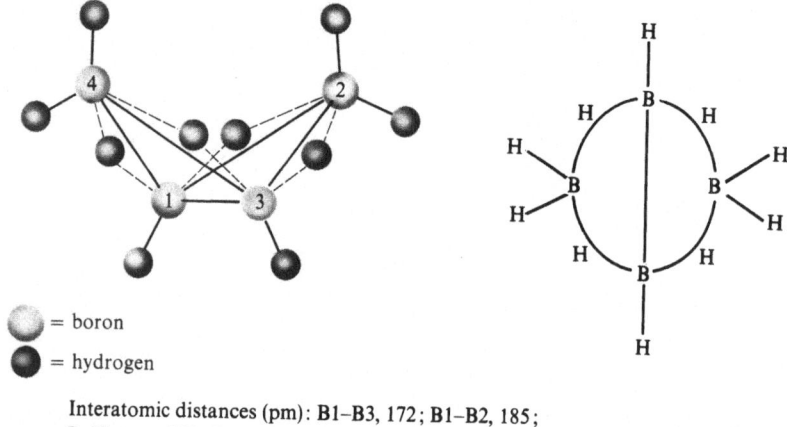

= boron

= hydrogen

Interatomic distances (pm): B1–B3, 172; B1–B2, 185; B–H$_{terminal}$, 119; B1–H$_\mu$, 133; B2–H$_\mu$, 143

Fig. 3–3   Structure and representation of the bonding of tetraborane(10), B_4H_{10}

(Fig. 3–3)[3] for example, there are 22 electrons $(4 \times 3 + 10)$ available for bonding. Six pairs are allocated to the six terminal B–H bonds, four pairs are allocated to the four BHB bridges, and the remaining two electrons are allocated to a two-centre electron-pair bond between borons 1 and 3, which are separated by an appropriate interatomic distance. In this compound each boron is four coordinate and assumed to be sp^3 hybridized, with the hybrid orbitals pointing towards the neighbouring atoms. The boron–hydrogen distances shown in Fig. 3–3 are those determined by electron-diffraction. As with diborane, the positions of the hydrogens determined by X-ray crystallography are less reliable.

The two types of boron atom in this molecule are clearly distinguished in its ^{11}B n.m.r. spectrum (Fig. 3–4).[4] Borons 2 and 4 have an environment similar to that of the boron atoms of diborane, in that each is attached

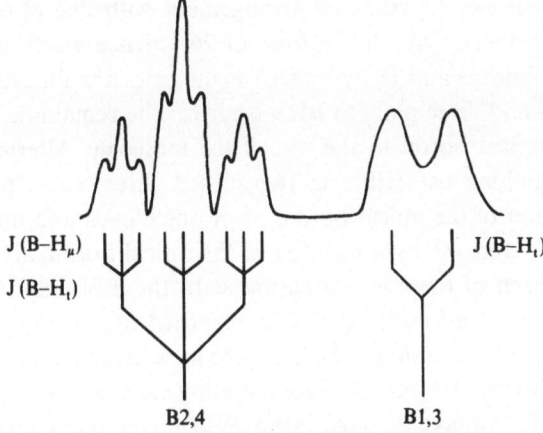

Fig. 3–4 ^{11}B n.m.r. spectrum of tetraborane(10), B_4H_{10}

to two terminal and two bridging hydrogens, and the resonance arising from them consists of a triplet of triplets. Borons 1 and 3 each have only one terminal hydrogen attached, and give rise to a doublet at somewhat higher field, with less well-defined fine structure. The integrated intensities of the large triplet and of the doublet are identical, showing that they are due to equal numbers of boron atoms.

In pentaborane(11), B_5H_{11} (Fig. 3–5), the boron skeleton is effectively

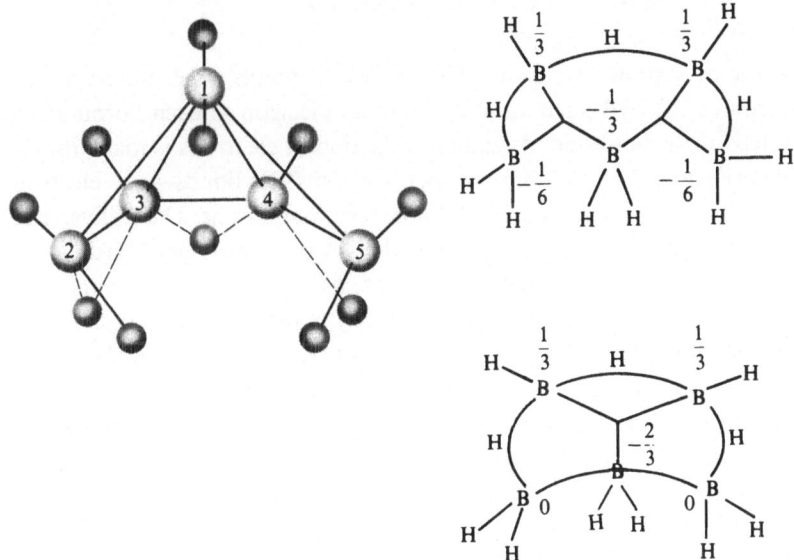

Interatomic distances (pm): B1–B2, 5, 187; B1–B3, 4, 172; B2(5)–B3(4), 176; B3–B4, 177

Fig. 3–5   Structure and representation of the bonding of pentaborane(11), B_5H_{11}, with approximate charge distributions

a pentagonal-based pyramidal arrangement with one of the pentagonal base sites unoccupied. Of the total of 26 valence electrons contributed by the five borons and 11 hydrogens, eight pairs are allocated to terminal B–H links, and three pairs to BHB bridges. The remaining four electrons bond the apical boron to the rest of the molecule. Alternative bonding schemes, making use either of two closed three-centre bonds and sp^3 hybridization of the apical boron, or of one closed and one open three-centre bond and sp^2 hybridization of the apical boron, are illustrated in Fig. 3–5. Each of these is compatible with the molecular structure, and neither representation alone can be regarded as a completely adequate description of the bonding, which may be looked on as a resonance hybrid to which these are the principal contributing canonical forms. Using them, it is possible to make a rough estimate of the charge distribution in the molecule assuming the pairs of bonding electrons in the various bridge bonds to be allocated as shown in Fig. 3–6. For example, in assessing the

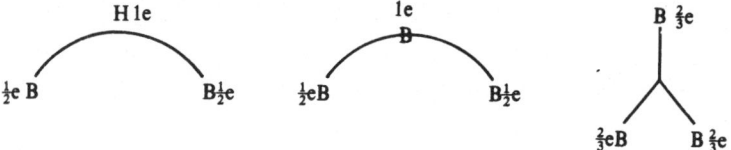

Fig. 3-6 Allocations of electrons in three-centre bonds that can be used to assess charge distributions

charges on boron atoms involved in BHB bonds, one allocates one electron to the hydrogen atom and half an electron to each boron atom involved in such a bond. A similar allocation of electrons is made rather arbitrarily to the three boron atoms of open BBB bonds—one electron to the central boron and half each to the terminal borons. The equivalence of the three boron atoms of a closed BBB bond makes two-thirds of an electron an appropriate allocation to each. At this level of approximation, hydrogen atoms are regarded as electrically neutral, so the two electrons of two-centre B–H bonds, like those of B–B bonds, are distributed equally, one apiece, between the two atoms thus linked. The charge on a particular boron atom is then deduced by adding up its share of the electrons in the four bonds it is involved in, and comparing the total with three, the number of electrons required for electrical neutrality.

Clearly, such arguments as these give only an approximate indication of the likely charge distribution in a molecule; they seriously exaggerate the differences between one boron and another, but at least provide a guide to qualitative differences. When they are applied to pentaborane(11), whichever canonical form in Fig. 3–5 is used, the apical boron is found to be more negatively charged than the remainder, a general characteristic of the higher boranes which are accordingly more susceptible to electrophilic attack at this point. That the apical boron is the most negatively charged in pentaborane(11) is supported by its [11]B n.m.r. spectrum,[5] which when recorded at a suitably high frequency (Fig. 3–7) resolves the signals due to the three types of boron into a triplet due to borons 2 and 5, each of which has two terminal hydrogens attached; a doublet due to borons 3 and 4, each of which has only one hydrogen attached; and a less intense *doublet* due to the apical boron, B1 (as if this boron had only *one* terminal hydrogen attached), at appreciably higher field than the other signals, indicating greater shielding of, i.e. a greater electron density

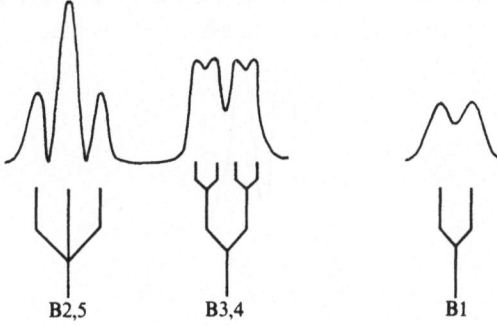

B2,5   B3,4   B1

Fig. 3–7 [11]B n.m.r. spectrum of pentaborane(11), B_5H_{11} (spectrum recorded at 70·58 MHz)

around, this boron. The doublet nature of this last signal arises because one of the two 'terminal' hydrogen atoms attached to B1 is close enough to borons 2 and 5 to be intermediate in character between bridging and terminal. Another interesting feature is the doublet fine structure of each component of the doublet due to borons 3 and 4, as if these were attached

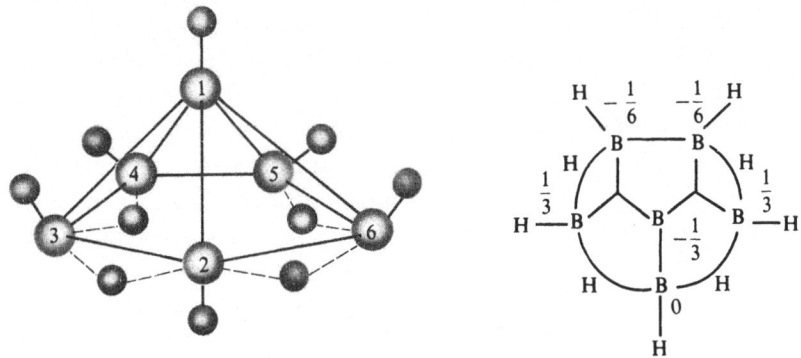

Interatomic distances (pm): B1–B2, 174; B1–B3,6, 175; B1–B4, 5, 180; B2–B3, 6, 179; B3(6)–B4(5), 174; B4–B5, 160

Fig. 3–8 Structure and representation of the bonding of hexaborane(10), B_6H_{10}

to only one bridging hydrogen apiece, for example, that between them. The remaining bridging hydrogens evidently exchange too rapidly at room temperature to give appropriate fine structure to the other peaks.

The framework bonding in the pentagonal pyramid of hexaborane(10), B_6H_{10} (Fig. 3–8), is most conveniently described in terms of two two-centre and two closed three-centre bonds. The basal two-centre link, only 160 pm in length, is one of the shortest boron–boron distances found in the boron hydrides, and differs markedly from the other basal boron–boron distances. Even though crystallographically there is this marked difference between the basal single bond and the remainder, in solution the bridging hydrogens can evidently exchange positions very readily, as shown by the [11]B n.m.r. spectrum (Fig. 3–9), in which all the basal boron atoms appear

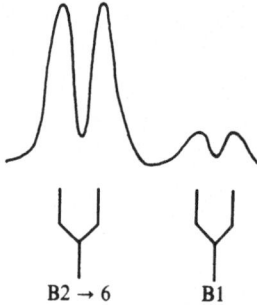

B2 → 6   B1

Fig. 3–9   [11]B n.m.r. spectrum of hexaborane(10), $B_6 10_{10}$

equivalent. Again, the apical boron absorbs at higher field than the basal borons, as is appropriate for a nucleus in a region of higher electronic charge.

The structure of decaborane(14), which has been studied by both X-ray diffraction[6] and neutron-diffraction,[7] is shown in Fig. 3–10. In this basket-shaped molecule the 10 boron atoms appear to occupy 10 of the 12 corners of an icosahedron (cf. Fig. 3–16 below), the two vacant corners being adjacent, though it is more helpful to regard the skeleton as based on an 11-cornered polyhedron with one corner vacant, as will be seen later. The four bridging hydrogens stitch up the open face. Average values for the seven distinct boron–boron interatomic distances and for the three distinct boron–hydrogen distances (these last are figures determined by neutron-diffraction) are given. Neutrons are scattered by atomic nuclei, not (like X-rays) by their attendant electron clouds, and the scattering power does not increase systematically with atomic number. Because protons scatter neutrons with an anomalous change of phase, so that hydrogen atoms appear as negative peaks in Fourier syntheses, and because [10]B nuclei have high absorption cross-sections for thermal neutrons, isotopically normal boron hydrides are unsuitable for investi-

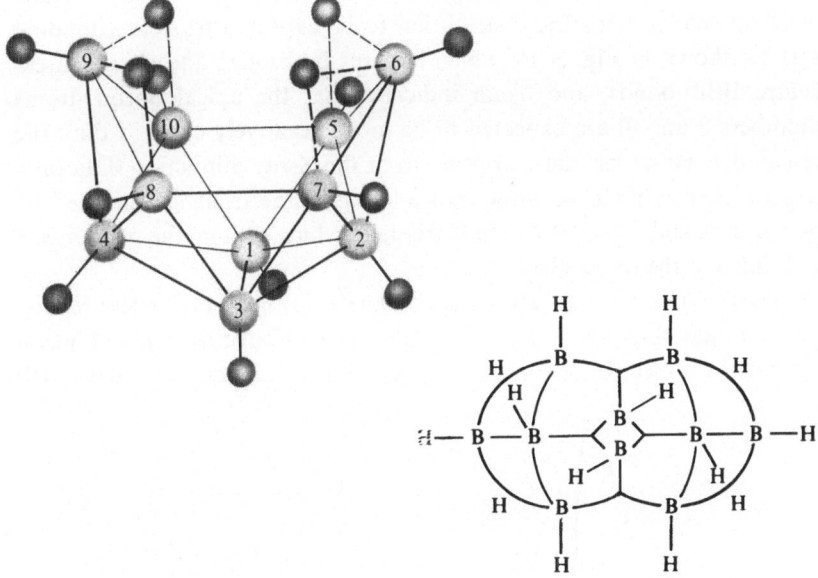

Boron–boron interatomic distances (pm):

	1–2,4 3–2,4	1–3	1–5,10 3–7,8	6–5,7 9–8,10	2–5,7 4–8,10	2–6 4–9	5–10 7–8
by X-ray diffraction[6]	179	171	177	177	178	172	201
by neutron- diffraction[7]	178	177	176	177	179	172	197

Boron–hydrogen interatomic distances[7] (pm):

terminal B–H; 118

bridging B–H; 130 (to borons 5, 7, 8, and 10); 135 (to borons 6 and 9).

Fig. 3–10   Structure and representation of the bonding of decaborane(14), $B_{10}H_{14}$

gation by neutron-diffraction. The nuclei ^2H and ^{11}B are satisfactory for neutron-diffraction work, however, and the study on decaborane(14) was carried out on a sample enriched in these isotopes, cooled to $-160°$ to minimize thermal motion of the nuclei.

In contrast to B–H distances determined by X-ray diffraction, those in Fig. 3–10 are comparable with the terminal and bridging B–H distances found in other substances by electron-diffraction. The BHB bridge bonds are unsymmetrical, with the bridging hydrogens slightly nearer to borons 5, 7, 8, and 10 than to borons 6 and 9.

Because the X-ray diffraction data reflect the electron density in decaborane(14), and the neutron-diffraction data the positions of the atomic nuclei, a comparison of the two gives some indication of the distribution of the bonding electrons in the molecule.[8] This shows quite good agreement with the distribution to be expected from the bonding scheme shown in Fig. 3–10, which features both open and closed three-centre BBB bonds, and again indicates that the apical boron atoms (numbers 2 and 4) are expected to be more negatively charged than the remainder. However, there appears from the X-ray diffraction data only slight evidence of closed three-centre bonds in the triangles enclosed by borons 1, 5, and 10 and 3, 7, and 8, triangles which contain the two longest B–B links in the molecule.

The ^{11}B n.m.r. spectrum of decaborane(14) (Fig. 3–11)[9] is less readily assigned than those of the lower hydrides. The four distinct types of boron atom in the molecule (borons 1 and 3; 2 and 4; 5, 7, 8, and 10; and 6 and 9)

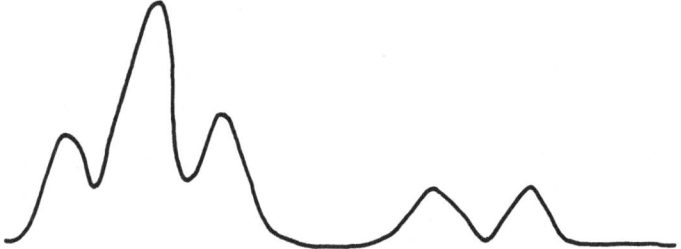

Fig. 3–11 ^{11}B n.m.r. spectrum of decaborane(14), $B_{10}H_{14}$

apparently do not differ enough to give the expected four doublets, one of intensity twice that of the others, as separately resolved signals. Indeed as a result of the overlapping of peaks, the spectrum superficially appears appropriate for a molecule containing terminal BH_2 and BH units (cf. Fig. 3–4).

The infrared spectra of the higher boranes, like diborane, contain absorptions in the region 2500–2620 cm^{-1} attributable to stretching vibrations of the terminal B–H groups (Table 3–1). Compounds containing terminal BH_2 groups (for example B_4H_{10}, B_5H_{11}) give strong doublet absorptions assignable to symmetric and antisymmetric vibrations. Compounds containing only terminal B–H groups (for example B_5H_9, B_6H_{10}, $B_{10}H_{14}$) give broad singlet, or closely spaced multiplet, absorptions the form of which is a consequence of the number of distinct B–H groups in the molecule. The absorptions due to stretching vibrations of BHB bridging groups (also listed in Table 3–1) are normally less intense and may occur over a wider frequency range (from below 1500 cm^{-1} to about 2200 cm^{-1}), within which the presence of combination bands makes assignment less certain, even when the spectra of deuterated samples are also available.

Relatively detailed treatments of the bonding in various boranes[10]

Table 3–1 Terminal and bridging B–H stretching absorptions in the infrared spectra of some boron hydrides

Compound	$\nu(B-H_{terminal})$ cm^{-1}	$\nu(B-H_{bridging})$ cm^{-1}
B_2H_6	2612vs, 2525vs	1882m, 1602vs
B_4H_{10}	2600vs, 2510vs	2160s, 1870w, 1710w, 1580w
B_5H_9	2610s, 2600s	1844w, 1802m
B_5H_{11}	2600vs, 2500vs	2050s, 1910w
B_6H_{10}	2610vs	1950s, 1490vs
$B_{10}H_{14}$	2565vs (broad)	1925m, 1885m

Relative intensities of absorptions are indicated as follows:
vs = very strong; s = strong; m = medium; w = weak

have allowed more reliable assessments of their electron distributions to be made than by the crude method outlined above. The following generalisations hold:

1. All atomic charges are between $+0.1$ and $-0.1e$ (the values obtained from the crude method outlined above greatly exaggerate charge differences).

2. Apical borons consistently are found to be more negatively charged than other borons.

3. The boron atoms of BH_2 groups are more positively charged than the borons of BH groups.

4. Terminal hydrogen atoms are invariably negatively charged, and bridging hydrogen atoms positively charged.

5. Hydrogen atoms acting as bridges between BH and BH_2 groups are more strongly bound to the BH group.

6. The electron density directly between two boron atoms that are hydrogen-bridged is very low.

The fourth of these generalizations incidentally confirms what was found to be the case with diborane, which is that if the absorption in the ^{1}H n.m.r. spectrum due to a bridging proton occurs at higher field than that due to a terminal proton, this cannot be taken as indicating that the former proton is in a region of greater overall negative charge.

3–3 Topological treatment of boron hydrides

The examples given above show that the structures of many boron hydrides are intelligible in valence bond terms without recourse to any more complicated concept than that of the three-centre bond linking sp^2- or sp^3-hybridized boron atoms, even in situations where certain boron atoms have as many as six near neighbours. A detailed consideration of the structures of all known boron hydrides led W. N. Lipscomb[1] to draw up certain generalizations about the topology of these structures, from which it was possible to make predictions about the likely existence of other as yet unknown hypothetical boranes.

In nearly all the boron hydrides known, every boron atom has at least one terminal hydrogen atom attached to it. If the formula of a particular hydride is written as B_pH_{p+q}, its structure may be regarded as composed of:

p B–H units each formally supplying two electrons (since two of the four valence electrons provided by one boron and one hydrogen atom are needed for the terminal B–H bond), and

q hydrogen atoms, each supplying one electron.

These may be regarded as held together by the following types of link:

s   three-centre BHB bonds;

t   three-centre BBB bonds (either open or closed);

y   two-centre B–B bonds; and

x   two-centre B–H bonds (if there are altogether **p+x** terminally attached hydrogens).

For these, the following relationships hold.

(1) A consideration of the number of hydrogen atoms shows that **q** must equal **s+x**.

(2) The total number of three-centre bonds **(s+t)** must equal the number of boron atoms **(p)**, since each boron provides four orbitals but only three electrons.

(3) A total of **p+q/2** electron pairs are used in **s+t+y+x** bonds; therefore **p+q/2** must equal **s+t+y+x**.

These relationships may conveniently be expressed in the following **equations of balance**:

$$x = q - s \tag{3-1}$$

$$t = p - s \tag{3-2}$$

$$y = \tfrac{1}{2}(s - x) \tag{3-3}$$

Using these equations of balance, possible values of **s**, **t**, **y**, and **x** can be computed for known and hypothetical boranes. For example, consider how diborane, B_2H_6, for which **p** = 2 and **q** = 4, can be built up from two B–H units and four hydrogen atoms. As the molecule contains only two boron atoms, it can contain no three-centre BBB bonds: therefore **t** must be zero. From Eqn (3–2), it follows that **s** = 2, i.e. that there are two three-centre BHB bonds in diborane. Next, from Eqn (3–1), we see that **x** = 4 − 2 = 2, i.e. that there are two terminally attached hydrogen atoms *additional to* those in the B–H units we started from. Finally, from Eqn (3–3), **y** = (2−2) = 0; there is no two-centre B–B bond in the molecule. We conclude that the structure in which two BH_2 units are held together by two three-centre BHB bonds is the only structure for diborane compatible with the equations of balance. This structure is referred to as having (2002) topology, the figures in parentheses being the values of **s, t, y,** and **x** in that order.

As a check, one can consider the possibilities that **s,** the number of BHB bonds, might have values of 0, 1, 3, or 4. Higher values are excluded since **q** defines the upper limit on possible values of **s.**

If **s** = 0, then from Eqn (3–2), **t** = 2; from Eqn (3–1), **x** = 4; and from Eqn (3–3), **y** = −2.

If **s** = 1, then from Eqn (3–2), **t** = 1; from Eqn (3–1), **x** = 3; and from Eqn (3–3), **y** = −1.

If **s** = 3, then from Eqn (3–2), **t** = −1; from Eqn (3–1), **x** = 1; and from Eqn (3–3), **y** = 1.

If **s** = 4, then from Eqn (3–2), **t** = −2; from Eqn (3–1), **x** = 0; and from Eqn (3–3), **y** = 2.

In all four cases, a negative value is obtained for one or other of the parameters, a result showing that not one of these sets of **s, t, y, x** numbers can have physical significance.

The same process, of trying values of **s** ranging from zero to **q,** can be applied to other boron hydrides. For tetraborane(10), B_4H_{10}, for example, one obtains the following sets of **s, t, y, x** numbers: 0, 4, −3, 6; 1, 3, −2, 5; 2, 2, −1, 4; 3, 1, 0, 3; 4, 0, 1, 2; 5, −1, 2, 1; and 6, −2, 3, 0. Of these the first three and last two sets of values can be excluded as they involve negative values of either **t** or **y.** The topology (4012) is that actually adopted by tetraborane(10) (Fig. 3–3). The topology (3103) appears to be a possible alternative, as shown below:

The hypothetical (3103)
structure for B_4H_{10}

The number of possible sets of **s, t, y, x** values increases as **p** and **q** increase, yet relatively few of the many hypothetical hydrides are known, and isomeric forms of a particular hydride are not normally isolable. Some further principles can be stated that eliminate many hypothetical formulations, including for example the (3103) structure for tetraborane(10) shown above.

(1) All known boron hydrides have at least a two-fold element of symmetry, so it is assumed that any new hydride probably would have at least a plane, centre, or two-fold axis of symmetry. Low symmetry appears to provide centres of reactivity. The (3103) structure for tetraborane(10) can be excluded because it lacks such an element of symmetry.

(2) No two adjacent boron atoms may be connected by both a closed and an open three-centre bond, apparently because of the large angular separation (90°) of the orbitals that would have to be used.

(3) Only one terminal hydrogen, and no bridging hydrogen, may be attached to a boron which is bound to five neighbouring borons. This restricts BHB bridges and BH_2 groups to the edges of boron frameworks.

(4) If a boron is bound to four other boron atoms, it will probably not make use of more than one BHB bridge.

(5) A boron atom that is bound to only two other boron atoms will be involved in at least one BHB bridge.

Such empirically-derived generalizations as these have been used by Lipscomb to consider the possible existence of hypothetical hydrides. Values of the **s, t, y, x** numbers are calculated for particular combinations of **p** and **q**; sets with negative values for one or other of the parameters are eliminated; the remaining sets are written out as atomic networks that fit the above rules. Throughout, one underlying assumption is that *each boron atom provides four orbitals, and therefore is involved in precisely four bonds*, whether these be two-centre or three-centre bonds.

To illustrate the application of this method to hypothetical species, one can consider the possible existence of triborane(9), B_3H_9, which is believed to be one of the initial products of thermal decomposition of diborane. Trying values of **s** in the range 0–6, one obtains only one set of **styx** numbers (3003) that does not contain a negative value for either **t** or **y**. A satisfactory structure for this species has a triangular arrangement of three BH_2 groups linked by hydrogen bridges along the edges of the B_3 triangle, as shown in Fig. 3–12a.

Another species believed to be involved in the thermal decomposition of diborane is triborane(7), B_3H_7. For this, **styx** values of either 2102 or 3011 fit the equations of balance. A structure with the former topology is shown in Fig. 3–12b; it has the required symmetry, but is considered somewhat unsatisfactory because of unfavourable bond angles at the boron atoms. The bond angles required for a 3011 topology (Fig. 3–12c) also appear unsatisfactory. The other possible structures shown for B_3H_7 (Fig. 3–12d–f) are unsatisfactory in that they make use of only three orbitals on one of the borons. A plausible (2202) structure for the hypothetical tetraborane(8) is shown in Fig. 3–12g. The reader interested in further examples will find many in Lipscomb's book.

The formulae and **styx** numbers of boron hydrides that have been

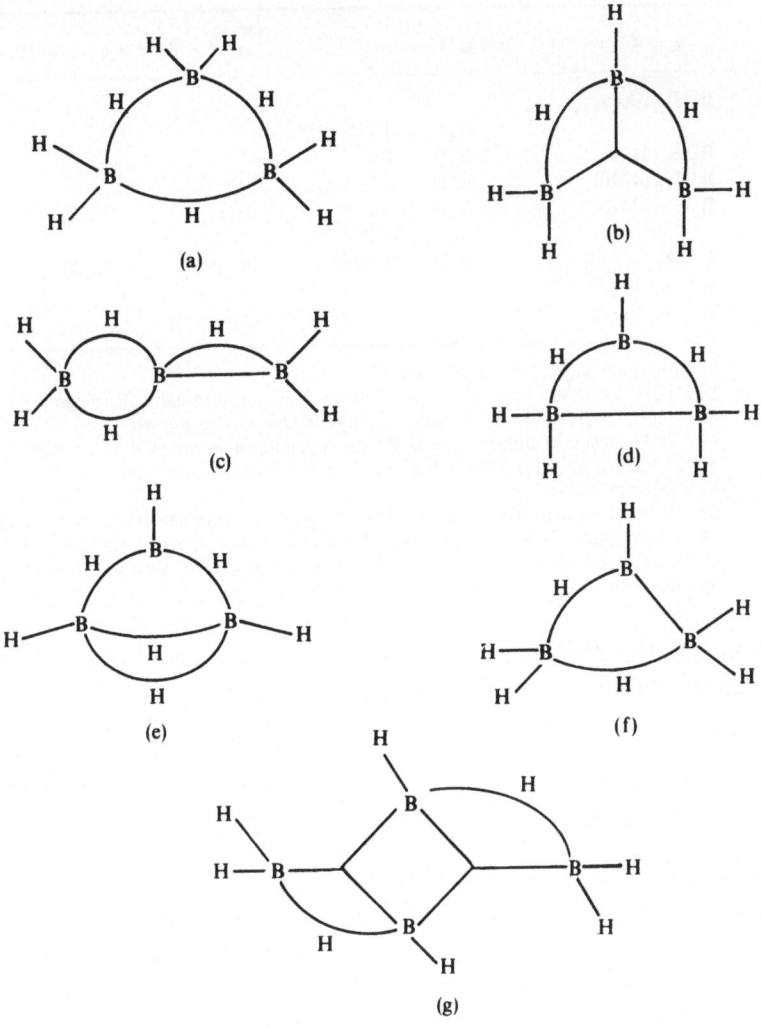

Fig. 3-12   Possible structures for species $B_3H_9(a)$, B_3H_7 (b → f), and $B_4H_8(g)$.

characterized well enough to indicate their topologies are listed in Table 3-2. A few of these hydrides ($B_{10}H_{16}$, $B_{16}H_{20}$, $B_{18}H_{22}$, and $B_{20}H_{16}$) have **styx** numbers that do not satisfy the equations of balance already given. This is because in these compounds not all the boron atoms have at least one terminal hydrogen atom attached, one of the conditions from which the equations of balance were derived. For molecules in which r boron atoms have no terminal hydrogen attached, which may be written $B_{p+r}H_{p+q}$ in order to preserve the condition that they contain p B–H

Table 3-2 **Formulae and styx numbers of known neutral boron hydrides** $B_{p+r}H_{p+q}$

$q-r = 4$	$q-r = 6$	$q-r = 8$	$q-r = 10$
B_2H_6 (2002)			
	B_4H_{10} (4012)		
B_5H_9 (4120)	B_5H_{11} (3203)		
B_6H_{10} (4220)	B_6H_{12} (4212?)		
B_8H_{12} (4420)	B_8H_{14} (4412?)	B_8H_{16}	$^aB_8H_{18}$ (8032)
	$^bB_9H_{15}$ (5421)		
$B_{10}H_{14}$ (4620)	$^cB_{10}H_{16}$ (8250)*	$^dB_{10}H_{18}$	$B_{10}H_{20}$
$B_{16}H_{20}$ (6, 10, 4, 0)*			
$^eB_{18}H_{22}$ (6, 12, 4, 0)*			

A compound $B_{20}H_{16}$ is also known.
^{a} Two B_4H_{10} residues linked by a B–B link replacing a terminal B–H of each.
^{b} An isomeric B_9H_{15}, probably with topology (6330), is also known.
^{c} Two B_5H_9 residues linked by a B–B link replacing a terminal B–H of each. One isomer has the apical borons linked; another has basal borons linked.
^{d} Two isomers known.
^{e} Two isomers known; each consists of two $B_{10}H_{14}$ residues sharing borons 5 and 6.
* These compounds have $r \neq 0$, i.e. they contain some boron atoms with no terminal hydrogen attached; modified equations of balance need to be used to arrive at their styx numbers.

groups, a consideration of the electron, orbital, and hydrogen atom balances leads to the modified equations of balance:

$$x = q-s \tag{3-4}$$
$$t = p+r-s \tag{3-5}$$
$$y = \tfrac{1}{2}(s+r-x) \tag{3-6}$$

which reduce to the earlier equations when $r = 0$. For example, the compound $B_{10}H_{16}$ consists of two pentaborane(9) residues linked by a two-centre B–B bond replacing a terminal B–H of each, and so has $p = 8$, $q = 8$, $r = 2$, $s = 8$, $t = 2$, $y = 5$, and $x = 0$.

Such treatments of the topologies of the boron hydrides are not limited to neutral species. For charged species, written $B_{p+r}H_{p+q+c}^{c+}$, the equations of balance need further modification to allow for the fact that these species contain c fewer electrons than the neutral species. The generalized equations of balance then become

$$x = q+c-s \tag{3-7}$$
$$t = p+r+c-s \tag{3-8}$$
$$y = \tfrac{1}{2}(s+r-3c-x) \tag{3-9}$$

These too reduce to those given earlier if r and c equal zero.

Examples of known borane anions are the species BH_4^-, $B_2H_7^-$, and $B_3H_8^-$. Application of the generalized equations of balance to these species, none of which contain boron atoms with no hydrogen attached ($r = 0$), leads to the following **styx** numbers:

$BH_4^- : p = 1, q = 4, c = -1, s = 0, t = 0, y = 0,$ and $x = 3$

$B_2H_7^- : p = 2, q = 6, c = -1, s = 1, t = 0, y = 0,$ and $x = 4$

$B_3H_8^- : p = 3, q = 6, c = -1, s = 1, t = 1, y = 0,$ and $x = 4$

or $p = 3, q = 6, c = -1, s = 2, t = 0, y = 1,$ and $x = 3$

These structures are shown in Fig. 3–13. The (2013) topology for the $B_3H_8^-$ anion is that found by X-ray crystallographic studies[11] on

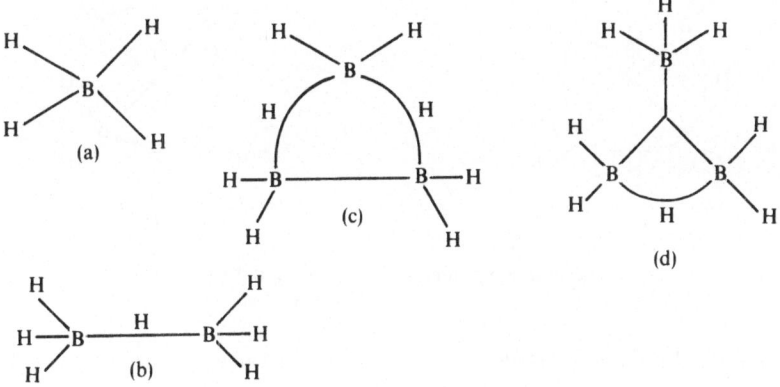

Fig. 3–13   Bonding representations of the anions BH_4^-, $B_2H_7^-$, and $B_3H_8^-$.

$(NH_3)_2BH_2^+B_3H_8^-$. The alternative (1104) topology (Fig. 3–13d) is unsatisfactory because it violates rule (5) above, though it represents an intermediate through which rearrangements of the hydrogen positions may occur.

3–4   The structure and bonding of pentaborane(9), B_5H_9

One of the lower hydrides listed in Table 3–1, pentaborane(9), B_5H_9, has a structure (Fig. 3–14) which reveals the limitations of the three-centre bond approach and the advantages of a partial molecular orbital treatment of molecules with relatively high symmetry. Since borons 2, 3, 4, and 5 are equivalent, a single (4120) formulation of the bonding as shown in Fig. 3–14a is unsatisfactory, as the apical boron (B1) forms two types of bond to the basal borons, two two-centre bonds using sp^2 hybrid orbitals to attach it to B2 and B4, and one open three-centre bond, to which it contributes a p orbital, to attach it to B3 and B5. Resonance between such a canonical form and its counterpart (Fig. 3–14b) needs to be invoked to render equivalent all the links between B1 and the base.

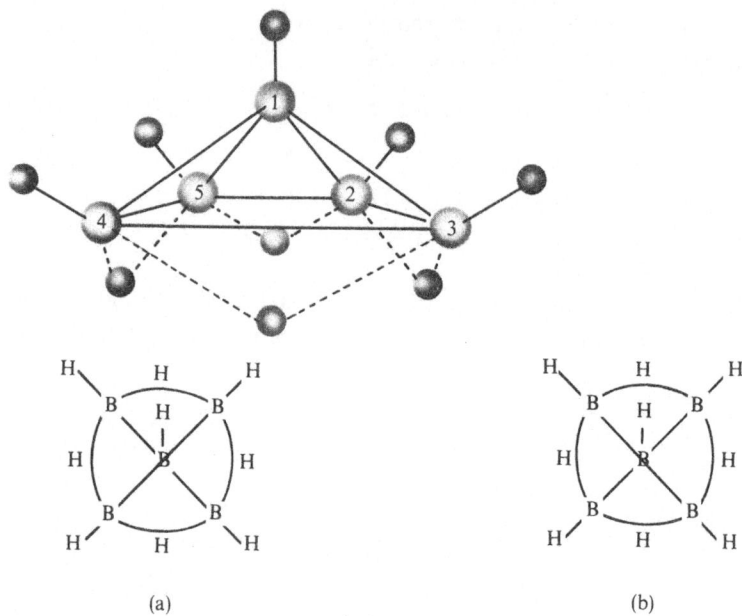

(a)   (b)

Interatomic distances: B1–B2 (etc.), 169 pm;
                        B2–B3 (etc.), 180 pm.

Fig. 3–14   The C_{4v} structure and representations of the bonding of pentaborane(9), B_5H_9

A better description follows from a partial molecular orbital treatment of the framework bonding. When the bonds to bridging and terminal hydrogens have been allowed for, each basal boron can contribute one sp^3 hybrid orbital pointing towards the four-fold axis of the molecule; the apical boron can contribute three atomic orbitals, an sp hybrid orbital pointing along the four-fold axis towards the centre of the base, and two p-orbitals at right angles to this. The three bonding orbitals that result from their interaction are shown in Fig. 3–15, which also shows, in the correlation diagram, that the remaining four orbitals are all anti-bonding.

The ^{11}B n.m.r. spectrum of pentaborane(9) qualitatively resembles that of hexaborane(10), the apical boron giving rise to a doublet at higher field than that due to the basal borons. The peak areas are, as expected, 1 : 4. The 'symmetrical top' symmetry of this molecule made it suitable for structural study by **microwave spectroscopy**. The microwave region of the spectrum is that in which radiation (of wavelength in the range 0·1–10 cm, i.e. frequency 3000–300 000 MHz) is of appropriate energy to excite directly transitions between rotational energy levels of molecules, transitions which otherwise can be studied by examination of the rotational fine structure of vibrational bands in the infrared region. Although only

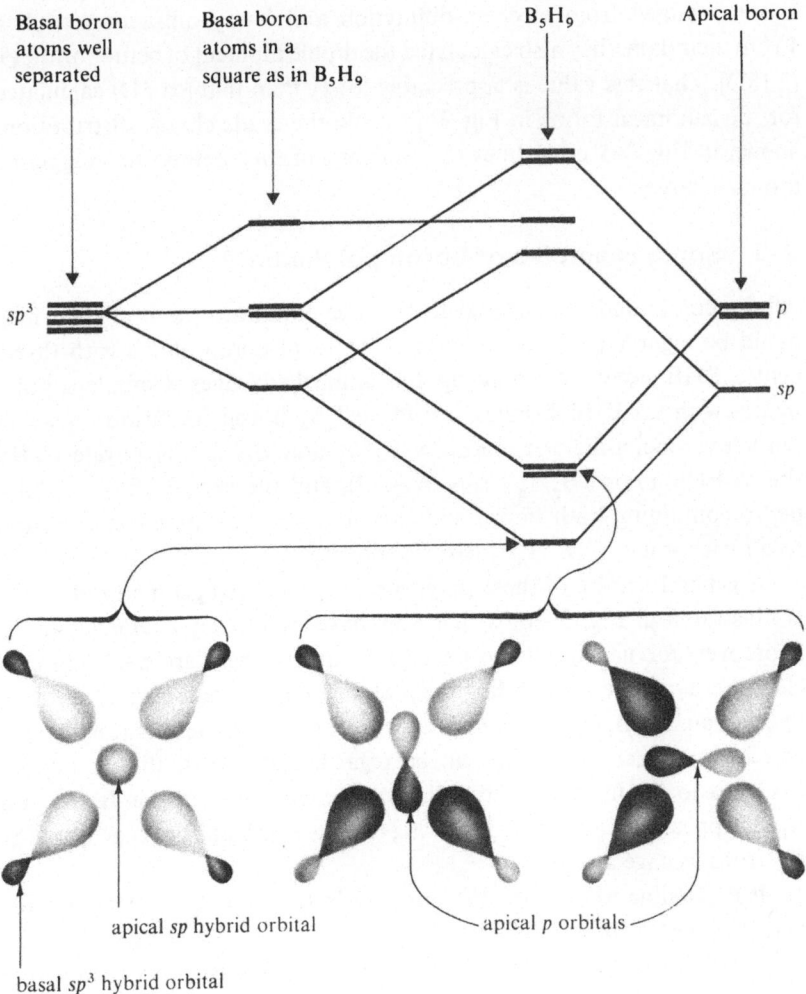

Basal boron atoms well separated

Basal boron atoms in a square as in B_5H_9

B_5H_9

Apical boron

sp^3

p

sp

apical sp hybrid orbital

apical p orbitals

basal sp^3 hybrid orbital

Fig. 3–15   The framework orbital correlation diagram and three bonding orbitals of pentaborane(9), B_5H_9

one moment of inertia of a symmetric top molecule can be determined— rotation about the top axis cannot be stimulated by absorption of radiation since there is no component of the dipole moment perpendicular to this axis—the molecular structure can nevertheless be determined if as many isotopic moments can be measured as there are independent bond distances and bond angles. Normal pentaborane(9) contains two isotopic symmetric-top species in comparative abundance—$^{11}B_5{}^1H_9$ and $^{10}B^{11}B_4{}^1H_9$ with the ^{10}B in the apical position— and by a study of their absorptions near 28 000 MHz, as well as those of the fully deuterated compound $B_5{}^2H_9$, H. J. Hrostowski and R. J. Myers[12] calculated apex– base and base–base distances identical within experimental error with those obtained from electron-diffraction and X-ray diffraction studies. From their data, they also calculated the dipole moment of pentaborane (9) (2·13D). That this value is appreciably lower than that (ca 5D) calculated for the canonical forms in Fig. 3–14 using the crude charge distributions shown in Fig. 3–5 underlines the tendency of this method to exaggerate atomic charges.

3–5 Some examples of boron polyhedra[13]

All the compounds considered so far have contained frameworks which could be regarded as incomplete polyhedra of boron atoms with three-centre BHB bonds stitching up the unfinished edges. Complete polyhedra, with no BHB bridges, are formed by boron in various types of derivative, including the chlorides B_4Cl_4 and B_8Cl_8, the boride CaB_6, the hydride anions $B_pH_p^{2-}$ ($p = 6 \rightarrow 12$), and the carbide $B_{12}C_3$. Polyhedra containing both boron and carbon atoms are moreover common to all carboranes $C_2B_nH_{n+2}$ (see Chapter 5).

A general feature of these polyhedra, the idealized geometry of which is illustrated in Fig. 3–16, is that they have exclusively triangular faces. Moreover, for nearly all the examples known, there are $p+1$ pairs of electrons available for bonding the p atoms of the cage together, as in the hydride anions $B_pH_p^{2-}$, in which each boron has one terminal hydrogen attached, so each B–H unit can be regarded as contributing a pair of electrons to the framework bonding. Some chlorides such as B_4Cl_4 and B_8Cl_8 appear exceptional in requiring only as many pairs of cage bonding electrons as cage atoms.

It is possible to describe the bonding in these cages in terms of two- and three-centre boron–boron bonds of the types already described, but it is rarely possible to draw a single canonical form that reflects the molecular symmetry and alone provides a realistic guide to the cage bonding. As was necessary for the two- and three-centre bonded description of pentaborane(9), one normally has to invoke resonance between a number of canonical forms, a number that increases with the size of the cage, so that there are for example some 70 plausible equivalent descriptions of the icosahedral cage bonding in the dodecahydro*closo*-dodecaborane(2−) anion, $B_{12}H_{12}^{2-}$. (The prefix '*closo*' is used in naming compounds in which the central core consists of a closed cage of boron atoms with no bridging hydrogens). Moreover, in relatively few cases does this approach lead to an understanding of why $p+1$ pairs of electrons are normally required to hold the cage together. This last point may be illustrated by considering whether, arguing solely from the premises that:

(a) only two-centre and/or three-centre boron–boron bonds are needed to describe the cage bonding; and that

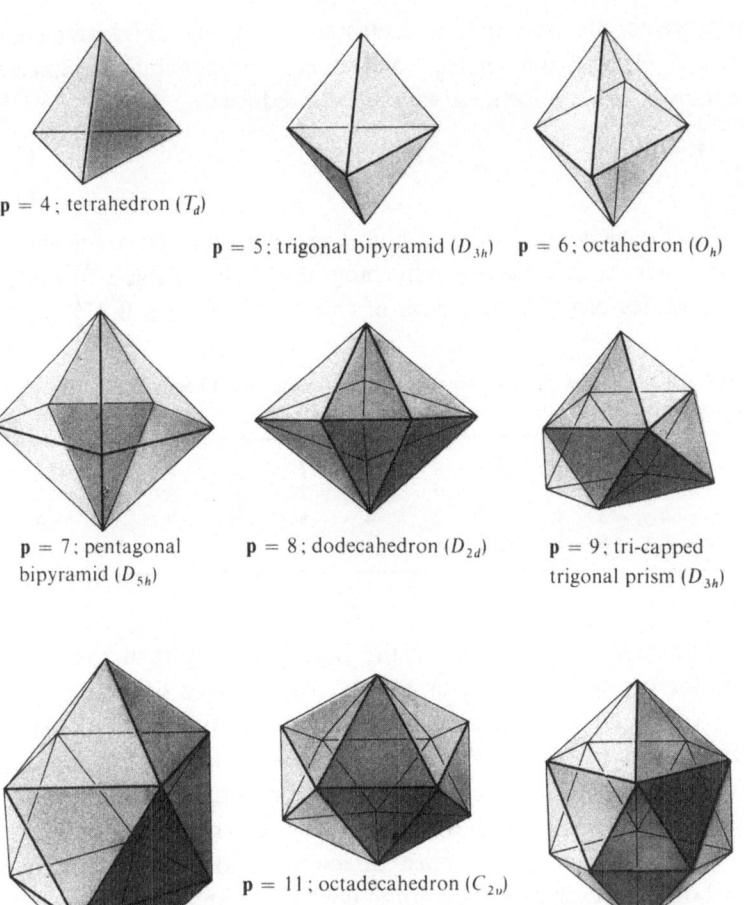

p = 4; tetrahedron (T_d)

p = 5; trigonal bipyramid (D_{3h})   p = 6; octahedron (O_h)

p = 7; pentagonal bipyramid (D_{5h})   p = 8; dodecahedron (D_{2d})   p = 9; tri-capped trigonal prism (D_{3h})

p = 11; octadecahedron (C_{2v})

p = 12; icosahedron (I_h)

p = 10; bi-capped Archimedean antiprism (D_{4d})

Fig. 3–16 Polyhedral structures adopted by *closo*-borane anions $B_pH_p^{2-}$ (p = 5-12) or neutral chloroboranes B_pCl_p (p = 4 or 8). The structures of *all* known higher boranes B_pH_{p+q} (and related species, such as carboranes – see chapter 6) may be regarded as based on the above 5→12 – cornered polyhedra, with all, all but one, or all but two corners occupied by skeletal atoms when q = 2, 4, or 6 respectively.

(b) each boron atom will be involved in three such cage bonds, one can decide whether a species $B_pH_p^{c+}$ is likely to be neutral or carry a positive or negative charge c. Application of the equations of balance to such species with values of **p** in the range 4 → 12 shows that positive values of c can be excluded, as these lead to negative values of y, (y = $\frac{1}{2}(s+r-x-3c) = -\frac{3}{2}c$ when s, r, and x all equal zero). Odd values of c may be discounted as they relate to odd-electron species, and an ionic charge greater than four may be dismissed as unlikely. This leaves neutral species B_pH_p and ions $B_pH_p^{2-}$ and $B_pH_p^{4-}$ as possibilities. Since the equations of balance for these cage species reduce to

$$t = p + c \tag{3-10}$$

and $\quad y = -\tfrac{3}{2}c \tag{3-11}$

a physically meaningful value of both t and y is obtained whether $c = 0$, -2, or -4 for each value of p in the range $4 \rightarrow 12$ (see Table 3–3). Only for certain values of p do the values of t and y for anions $B_pH_p^{2-}$ appear

Table 3–3   Values of t (the number of three-centre BBB bonds) and y (the number of two-centre B–B bonds) for cage species $B_pH_p^{c+}$

	p	4	5	6	7	8	9	10	11	12
$c = 0$, $y = 0$	t	4	5	6	7	8	9	10	11	12
$c = -2$, $y = 3$	t	2	3	4	5	6	7	8	9	10
$c = -4$, $y = 6$	t	0	1	2	3	4	5	6	7	8

obviously more appropriate for the cage geometry than the y and t values for the species B_pH_p or $B_pH_p^{4-}$. An example of an apt set of t and y values for the doubly charged anion is provided by the species $B_5H_5^{2-}$ ($p = 5$, $t = 3$, $y = 3$) for which a reasonable structure appears to be that (Fig. 3–17a) in which a trigonal bipyramidal arrangement of the five boron atoms is held together by three two-centre bonds linking the equatorial atoms and three open three-centre BBB bonds linking the apical borons to each of the equatorial borons in turn. In many other cases it is difficult to see, using this extended valence-bond approach, why the neutral species B_pH_p and anion $B_pH_p^{4-}$ should not also be capable of existence. Plausible canonical forms for hypothetical neutral species include that using four closed three-centre bonds, one in each of the four faces of a tetrahedral molecule, B_4H_4, or the use of six open three-centre bonds along the 12 edges of the octahedron of the *closo*-hexaborane B_6H_6 (Fig. 3–17b). A symmetrical array of six two-centre and eight three-centre bonds that appears appropriate for a hypothetical icosahedral anion $B_{12}H_{12}^{4-}$ is shown in Fig. 3–17c. In short, this approach leads one to expect the existence of a much wider range of compounds than is known.

A much clearer indication of the bonding electron requirements of a particular cage is given by molecular orbital approaches which take account of the cage symmetries, along the lines already illustrated for diborane and pentaborane(9).

For example, the tetrahedral boron cage of tetraboron tetrachloride (tetrachloro*closo*tetraborane), B_4Cl_4 (Fig. 3–18)[14] is apparently held

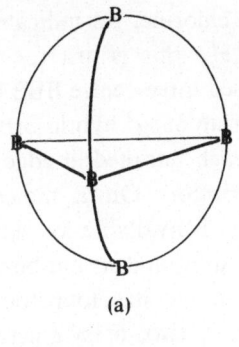

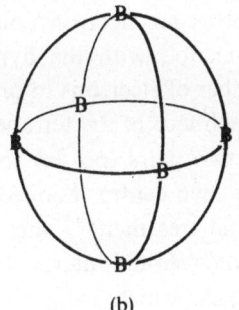

(a)   (b)

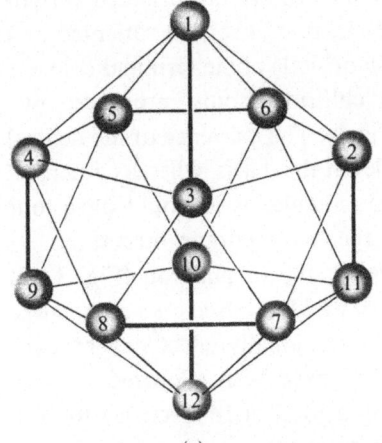

Two-centre bonds link the following pairs of atoms: 1,3; 5,6; 2,11; 4,9; 7,8; 10,12

Closed three-centre bonds occupy the following faces: 1,4,5; 1,2,6; 2,3,7; 3,4,8; 5,9,10; 6,10,11; 7,11,12; 8,9,12

(c)

Fig. 3–17 Plausible two- and/or three-centre bonding schemes for the cages of hypothetical species   (a) $B_5H_5^{2-}$ (b) B_6H_6 (c) $B_{12}H_{12}^{4-}$

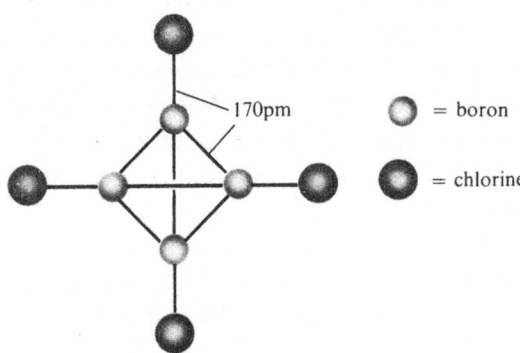

170pm

⚪ = boron

⚫ = chlorine

Fig. 3–18 The tetrahedral structure of tetrachloro*closo*tetraborane, B_4Cl_4

together by a total of eight electrons, assuming that each B–Cl link involves an electron apiece from boron and chlorine. As indicated in connection with the hypothetical hydride B_4H_4, this is just the right number of electrons to provide a pair for a closed three-centre BBB bond in each face of the tetrahedron, formed by overlap of sp^3 hybrid orbitals, but two pairs too few to provide a pair for each tetrahedral edge in a fully two-centre bonded valence-bond description. On a molecular orbital treatment,[15] the 12 boron atomic orbitals available for skeletal bonding should interact to give six formally bonding, and six antibonding orbitals, which suggests that the neutral molecule has four too few electrons. However, of the six bonding orbitals, two, of symmetry E, result solely from interactions between boron $2p$ orbitals oriented at right angles to the three-fold axes (the four B–Cl bond axes), i.e. oriented in a suitable manner to interact with filled $3p$ orbitals of the terminal chlorine atoms. 'Back coordination' from the chlorine atoms into these two orbitals of the B_4 tetrahedron appears likely. The existence of the neutral chloride B_4Cl_4 is accordingly intelligible on this basis, whereas a neutral hydride B_4H_4 appears less likely to be isolable, as it would have four electrons fewer than those required by the 'tetrahedrane' structure.

The next boron polyhedron illustrated, the octahedron (Fig. 3–19) occurs in the boride CaB_6, the structure of which is based on a giant cubic lattice of B_6^{2-} octahedra linked three-dimensionally effectively by two-centre electron-pair bonds. After electrons have been allocated to these bonds (six per octahedron), there remain a total of 14 electrons to hold the octahedron together. Each boron atom in the cage can conveniently be regarded as sp hybridized, like the apical boron of B_5H_9 in Fig. 3–15, with one sp hybrid orbital pointing outside the octahedron towards the next octahedron. The other sp hybrid orbital points towards the centre of the octahedron. The two p orbitals also available on each boron are then effectively surface orbitals in that their maximum extension is tangential to the (pseudo-spherical) octahedral surface. The seven bonding inter-actions between these orbitals have symmetries as illustrated in Fig. 3–19. The remaining 11 molecular orbitals have some antibonding character, and remain unoccupied. Similar arguments apply to the anion[16] $B_6H_6^{2-}$ the existence of which was accordingly predicted before any of its salts were known.

A corollary of this molecular orbital treatment is that because of the three-fold degeneracy of the highest occupied orbitals (the a_{1g} orbital is most unlikely to be that of highest energy), a neutral molecule B_6H_6 would be expected to have two unpaired electrons and be unlikely to be isolable, whereas on the three-centre bond extension of the valence-bond approach a plausible structure for this molecule can be drawn (Fig. 3–17b) with all electron spins paired.

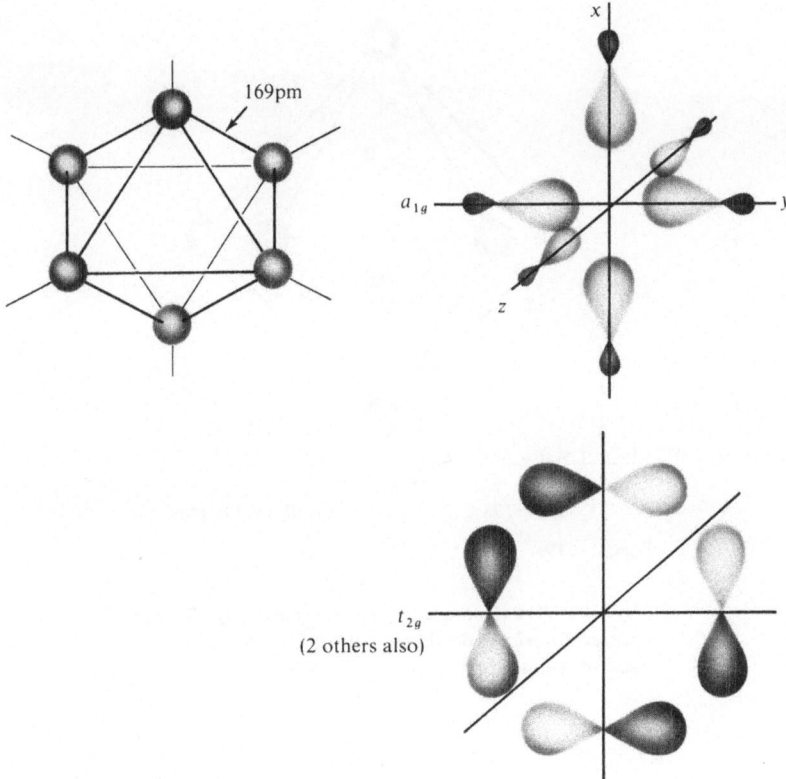

Fig. 3–19 The B_6 octahedron of $B_6H_6^{2-}$
and CaB_6, and the seven bonding orbitals

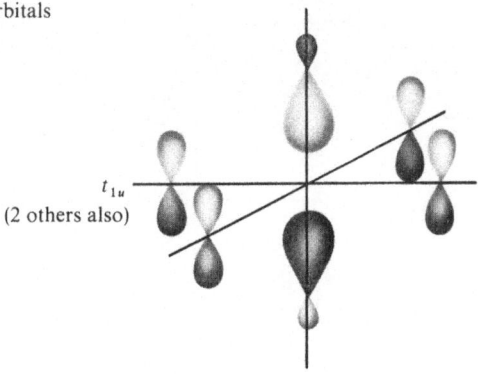

The structures of the larger boron cages depicted in Figs 3–20,[17] 3–21,[18] and 3–22[19] can be interpreted along similar lines. The atomic orbitals of the cage borons are conveniently considered in two sets; firstly, a set of radial orbitals, *sp* hybrid orbitals pointing towards the centre of the polyhedron; and secondly, a set of surface or tangential orbitals, orbitals which have their maximum extension in the pseudo-spherical surface of the polyhedron. The details of their interactions are too com-

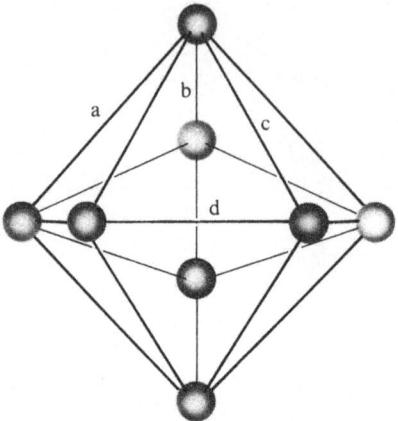

Mean B–B distances (pm):

	a	b	c	d	
B_8Cl_8	200	178	175	168	(mean B–Cl, 175 pm)[3]
$B_8H_8^{2-}$	193	174	178	158	(ref. 17)

Fig. 3–20 The dodecahedral (approximately D_{2d}) B_8 cage of octachloro*closo*- octaborane, B_8Cl_8, and of the octahydro*closo*-octaborate(2-) anion, $B_8H_8^{2-}$

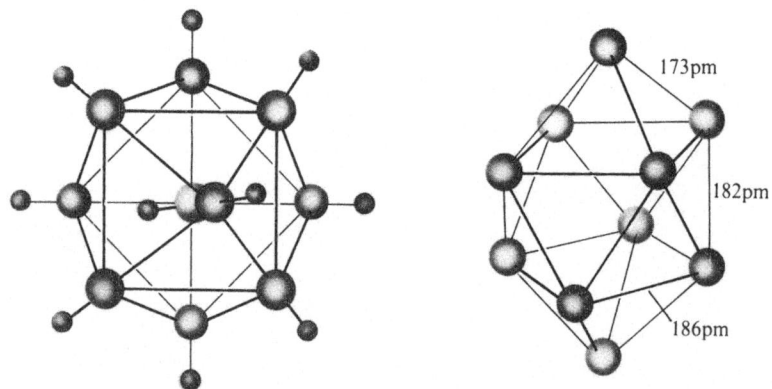

Fig. 3–21 The D_{4d} structure (bi-capped Archimedean antiprism) of the decahydro–*closo*-decaborate(2-) anion, $B_{10}H_{10}^{2-}$, with a second view of the B_{10} cage which illustrates the relationship to the square pyramidal cage of pentaborane(9), B_5H_9

plicated to be dealt with here, but again the number of orbitals likely to contribute to the bonding can be inferred from the structures. In the B_{12} icosahedron, for example (see Fig. 3–23), 13 molecular orbitals appear likely to be bonding orbitals out of the overall number of 36 that result from the interactions of the three atomic orbitals (one radial sp hybrid orbital and two tangential p orbitals) that each boron atom contributes.[20]

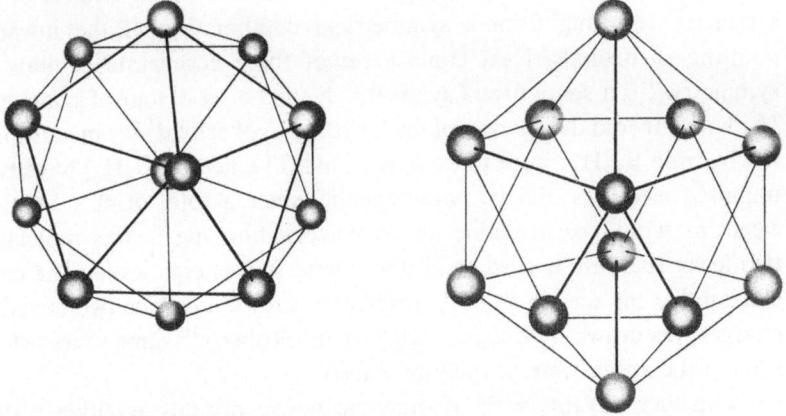

B–B interatomic distance along icosahedral edge, 177pm
B–B interatomic distance along five-fold axis (cage diameter), 334pm

Fig. 3–22 Two views of the icosahedral arrangement of 12 boron atoms that occurs in the dodecahydro*closo*dodecaborate(2−) anion, $B_{12}H_{12}^{2-}$, in boron carbide $B_{12}C_3$, and in elemental boron

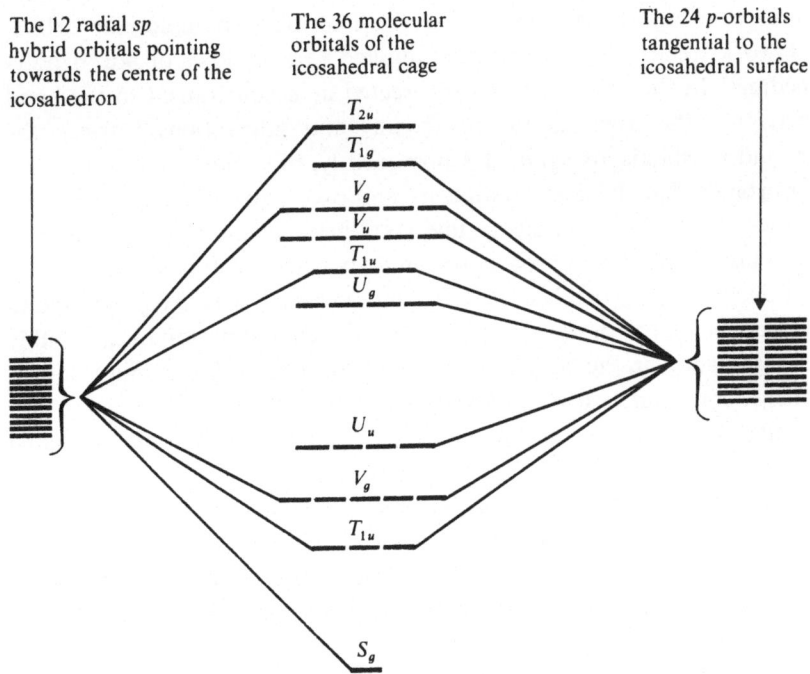

The 12 radial *sp* hybrid orbitals pointing towards the centre of the icosahedron

The 36 molecular orbitals of the icosahedral cage

The 24 *p*-orbitals tangential to the icosahedral surface

T_{2u}
T_{1g}
V_g
V_u
T_{1u}
U_g

U_u
V_g
T_{1u}

S_g

Fig. 3–23 The relative energy levels and degeneracies of the 13 bonding and 23 antibonding molecular orbitals of the icosahedral cage of the dodecahydro*closo*dodecaborate(2−) anion.[20] $B_{12}H_{12}^{2-}$

Of these 13, that of lowest energy is the unique bonding orbital of A_g symmetry resulting from a symmetrical combination of the inward-pointing sp orbitals. Next come a set of three degenerate orbitals of symmetry T_{1u}, a set of five of symmetry V_g, and a set of four of symmetry U_u. The four-fold degeneracy of the fourth type of orbital (U_u) means that, if a hydride $B_{12}H_{12}$ were to exist, it would (like neutral B_6H_6) have two unpaired electrons and be paramagnetic, since a total of $48 - 24 = 24$ electrons would be available for framework bonding in this molecule, two fewer than are needed to fill the lowest four energy levels. The conclusion that the species $B_{12}H_{12}$ would exist only if it carried two negative charges was drawn by Longuet-Higgins and Roberts[20] some years before salts of the anion were actually prepared.

A further advantage of treating the boron orbitals as either radial or tangential is that the distribution of electrons in the former will be sensitive to the inductive effect of substituents attached to the polyhedra, whereas the latter will be sensitive to mesomeric effects. The polyhedral boranes thus afford scope for the study of substituent effects over an extended delocalized system.

A rather different separation of molecular orbitals into two types—those that have no nodal surface containing the two-centre links (σ-orbitals) and those that have a nodal surface containing the two-centre links (π-orbitals)—is a feature of the free electron molecular orbital network treatment that has been applied to a number of boron poly-hedra.[21] In this, the electrons are treated as if constrained to the bond network—the edges of the polyhedra—and interactions between the σ- and π-orbitals are ignored. Consequently, this approach leads to sets of orbitals that differ in multiplicity and order from those predicted by molecular orbital treatments that use linear combinations of atomic orbitals. Using this approach, one finds that the neutral species such as tetrahedral B_4X_4, octahedral B_6H_6, and icosahedral $B_{12}H_{12}$ correspond to full occupancy of the σ-bonding orbitals. The formation of the doubly charged anion in the last two cases is intelligible only if the unique lowest energy π-orbital is also occupied, which appears reasonable since this orbital corresponds to the unique bonding combination of radial p-orbitals.

The dodecahedral cage of D_{2d} symmetry of eight boron atoms is known in the form of both the neutral chloride[3] B_8Cl_8 and the doubly charged anion[17] $B_8H_8^{2-}$ (see Fig. 3–20). That the two extra cage electrons possessed by the latter species are indeed *bonding* electrons is supported by the respective interatomic distances for B_8Cl_8 and $B_8H_8^{2-}$. The boron–boron distances in the latter are shorter than in the former. Detailed interpretation of the differences between the interatomic distances is not justified, because the values given represent average values for the four unique distances for a species of idealized D_{2d} symmetry; in both cases the actual cages are somewhat distorted from this.

To summarize: a useful picture of the distribution of the valence electrons in the higher boron hydrides is given by a valence-bond treatment using three-centre as well as two-centre bonds. A similar treatment of the polyhedral cages formed by boron is also possible, but the bonding picture it gives becomes increasingly complicated as the size of the cage increases. Large numbers of equivalent canonical forms have to be reckoned as contributing to the resonance hybrid, and it is not at all easy to discern how many valence-electrons will be needed to hold a particular cage together. On the other hand, molecular orbital treatments taking into account the symmetry of complete molecules allow the energy sequence, degeneracy, and bonding or antibonding properties of the molecular orbitals to be deduced, and so the number of bonding electrons needed can be assessed. As this tends to be at the expense of obscuring the approximate bond orders by which adjacent pairs of atoms are linked, the molecular orbital treatment is best used for symmetrical cages in which the symmetry alone means either that all the two-centre links are identical as in Figs 3–18, 3–19, and 3–22, or that they fall into a small number of symmetry-related types, as in Figs 3–20 and 3–21.

If one compares the skeletal structures of neutral boranes B_pH_{p+q} (Figs 3–3, 3–5, 3–8, 3–10, and 3–14) with the triangular-faced polyhedra shown in Fig. 3–16, not only are the former seen to be fragments of the latter, but two distinct types of skeletal structure can be discerned. For compounds with $q = 4$ (e.g., B_5H_9, B_6H_{10}, and $B_{10}H_{14}$), the skeletal boron atoms occupy all but one of the corners of the $(p+1)$-cornered polyhedron (i.e., of the octahedron, pentagonal bipyramid, and octadecahedron respectively). For compounds with $q = 6$ (e.g., B_4H_{10} and B_5H_{11}) the skeletal boron atoms occupy all but *two* of the corners of the $(p+2)$-cornered polyhedron (i.e., of the octahedron and pentagonal bipyramid respectively). In fact, the polyhedra shown in Fig. 3–16 form the basis for the structures of all known higher boranes, borane anions, and carboranes.[22,23] This point is discussed further in Chapter 6.

References

1. Lipscomb, W. N., *Adv. Inorg. Chem. Radiochem.*, 1959, 1, 117; Lipscomb, W. N., *Boron Hydrides*, W. A. Benjamin, New York, 1963. Specific references to earlier work will be found in these sources.
2. Rules for naming boron compounds are set out in *Inorg. Chem.*, 1968, 7, 1945.
3. Pawley, G. S., *Acta Cryst.*, 1966, 20, 631.
4. Hopkins, R. C., J. D. Baldeschwieler, R. Schaeffer, F. N. Tebbe, and A. Norman, *J. Chem. Phys.*, 1965, 43, 975.
5. Williams, R. E., F. J. Gerhart, and E. Pier, *Inorg. Chem.*, 1965, 4, 1239; Rietz, R. R., R. Schaeffer, and L. G. Sneddon, *J. Amer. Chem. Soc.*, 1970, 92, 3514.

6. Kasper, J. S., C. M. Lucht, and D. Harker, *Acta Cryst.*, 1950, **3**, 436.
7. Tippe, A., and W. C. Hamilton, *Inorg. Chem.*, 1969, **8**, 464.
8. Brill, R., H. Dietrich, and H. Dierks, *Angew. Chem. Internat. Edn.*, 1970, **9**, 524.
9. Williams, R. L., N. N. Greenwood, and J. H. Morris, *Spectrochim. Acta*, 1965, **21**, 1579; MacLean, D. B., J. D. Odom, and R. Schaeffer, *Inorg. Chem.*, 1968, **7**, 408.
10. Switkes, E., I. R. Epstein, J. A. Tossell, R. M. Stevens, and W. N. Lipscomb, *J. Amer. Chem. Soc.*, 1970, **92**, 3837; Switkes, E., W. N. Lipscomb, and M. D. Newton, *ibid.*, p. 3847.
11. Peters, C. R., and C. E. Nordman, *J. Amer. Chem. Soc.*, 1960, **82**, 5758.
12. Hrostowski, H. J., and R. J. Myers, *J. Chem. Phys.*, 1954, **22**, 262; see also Cohen, E. A., and R. A. Beaudet, *ibid.*, 1968, **48**, 1220.
13. Muetterties, E. L., and W. H. Knoth, *Polyhedral Boranes*, Marcel Dekker, New York, 1968.
14. Atoji, M., and W. N. Lipscomb, *Acta Cryst.*, 1953, **6**, 547; *J. Chem. Phys.*, 1953, **21**, 172.
15. Longuet-Higgins, H. C., *Quart. Rev.*, 1957, **11**, 121.
16. Schaeffer, R., Q. Johnson, and G. S. Smith, *Inorg. Chem.*, 1965, **4**, 917.
17. Guggenberger, L. J., *Inorg. Chem.*, 1969, **8**, 2771.
18. Dobrott, R. D., and W. N. Lipscomb, *J. Chem. Phys.*, 1962, **37**, 1779.
19. Wunderlich, J. A., and W. N. Lipscomb, *J. Amer. Chem. Soc.*, 1960, **82**, 4427.
20. Longuet-Higgins, H. C., and M. de V. Roberts, *Proc. Roy. Soc. (London)*, 1955, **A230**, 110.
21. Waddington, T. C., *Trans. Faraday Soc.*, 1967, **63**, 1313.
22. Williams, R. E., *Inorg. Chem.*, 1971, **10**, 210.
23. Wade, K., *Chem. Comm.*, 1971, 792.

General references

Refs 1, 13, and 15 (above) with the following:
Adams, R. M. (ed.), *Boron, Metalloboron Compounds and Boranes*, John Wiley (Interscience), New York, 1964.
Hawthorne, M. F., *Endeavour*, 1966, **25**, 146.
Muetterties, E. L. (ed.), *The Chemistry of Boron and its Compounds*, John Wiley, New York, 1967.
Eaton, G. R., and W. N. Lipscomb, *N.M.R. Studies of Boron Hydrides and Related Compounds*, W. A. Benjamin, New York, 1969.
Brotherton, R. J., and H. Steinberg (eds.), *Progress in Boron Chemistry. Vol. II*, Pergamon Press, Oxford, 1970.

Problems

3–1 Using the electron contributions shown in Fig. 3–6, calculate the approximate charges on the boron atoms of:

    (a) the species shown in Fig. 3–12;

    (b) the anions shown in Fig. 3–13;

    (c) pentaborane(9) (Fig. 3–14);

    (d) decaborane(14) (Fig. 3–10).

3–2 By a consideration of the electron, orbital, and hydrogen atom balances for

    (a) neutral species $B_{p+r}H_{p+q}$ and

    (b) charged species $[B_{p+r}H_{p+q+c}]^{c+}$

derive the following equations of balance

    (a) (i) $x = q-s$ (ii) $t = p+r-s$ (iii) $y = \frac{1}{2}(s+r-x)$

    (b) (i) $x = q+c-s$ (ii) $t = p+r+c-s$ (iii) $y = \frac{1}{2}(s+r-3c-x)$

in which p, q, r, s, t, y, x, and c are as defined in section 3–3.

3–3 By considering possible **styx** numbers for hypothetical species $B_2H_n^{c+}$ (where c can range from $+3$ to -3) determine for what values of c and n the appropriate equations of balance can be satisfied. Suggest structures for any species (apart from B_2H_6) that appear capable of existence. What principal absorptions are expected in their ^1H and ^{11}B n.m.r. spectra?

3-4   Discuss the possibility that the following hydrides might be capable of existence:

$$B_4H_6; B_4H_{12}; B_4H_{14}; B_5H_7; B_5H_8; B_5H_{13}$$

Suggest likely structures where appropriate.

3-5   Describe the principal absorptions that one might expect in the ^1H and ^{11}B n.m.r. spectra of the substances shown in Fig. 3–12 if these were sufficiently stable to be isolated and studied, and also those in Fig. 3–13.

3-6   (4220) Hexaborane(10) has a structure based on a pentagonal-based pyramidal arrangement of boron atoms (Fig. 3–8). What other values of s, t, y, and x are possible for this compound? Draw the structure, assuming the same disposition of boron atoms. Show how the same spatial arrangement of boron atoms can form the skeleton of plausible structures for isomers of hexaborane(12).

3-7   Discuss the advantages and disadvantages of describing· the bonding in boron hydrides in terms of two- and three-centre bonds,

3-8   Discuss the advantages and disadvantages of describing the bonding in boron hydrides in terms of molecular orbitals which are linear combinations of atomic orbitals of a form dictated by the symmetries of the complete molecules.

3-9   Discuss the relative merits of valence-bond and molecular orbital treatments of the bonding in boron hydrides.

4 The general chemistry of the boron hydrides

4-1 Introduction: some thermodynamic data

The boron hydrides in general are volatile, highly reactive substances. Their low boiling points (see Table 4–1) reflect their low molecular weights and feeble intermolecular forces. Their reactivities have a thermodynamic origin in that these compounds are markedly unstable with respect to replacement of their boron–boron and boron–hydrogen links by bonds to more electronegative elements. Some thermodynamic data bearing on this are in Tables 4–1 and 4–2.

Table 4–1 Physical and thermodynamic properties of some boron hydrides

	B_2H_6	B_4H_{10}	B_5H_9	B_5H_{11}	B_6H_{10}	$B_{10}H_{14}$
m.p. °C	−165·5	−120	−46	−122	−62	+99·7
b.p. °C	−92·5	16	60	63	108	213
ΔH_f° kJ/mole	36	66	73	103	95	31
ΔG_f° kJ/mole	87	—	175	201	—	216
ΔH_a kJ/mole	2400	4360	4700	5110	5460	8640
$\Sigma E(B–X)$ kJ/mole	2400	4360	4675	5110	5430	8450
ΔH_c kJ/mole	−1990	−3800	−4320	−4600	−5090	−8050
ΔH_c kJ/g	−72	−72	−69	−71	−68	−66

Thermodynamic data throughout relate to gaseous hydrides. ΔH_a is the enthalpy of atomization of the hydride in the gas phase. $\Sigma E(B–X)$ is the sum of the BH, BHB, BB, and BBB bond energy terms (see Table 4–2 for values) appropriate for the bonding representations in Figs 3–3, 5, 8, 10, and 14. ΔH_c is the enthalpy change for the reaction:

$$B_pH_{p+q} + \tfrac{1}{2}(4p+q)O_2 \rightarrow \tfrac{1}{2}pB_2O_{3(s)} + \tfrac{1}{2}(p+q)H_2O_{(g)}$$

Table 4–2 Some typical bond energy terms E(B–X) (kJ/mole)

X	H	B	C	N	O	F	S	Cl
E(B–X)	375	310	365	500*	540*	580*	400*	420*
E(BXB)	450	380						

*These values are minimal values; actual values vary widely from compound to compound, apparently with the degree of $B \rightleftharpoons X$ multiple bonding.

The positive values of the heat of formation ΔH_f° in Table 4–1 show that the volatile hydrides are *endothermic* with respect to elemental boron and hydrogen in their standard states. Moreover, the exclusively positive values of the free energy of formation, ΔG_f°, of the compounds in Table 4–1 show that they are thermodynamically unstable with respect to decomposition into boron and hydrogen.[1] They owe their existence to the absence of a suitable mechanism by which they can be converted into elemental boron and hydrogen at normal temperatures. All when heated strongly (700–800°C generally suffices) decompose to boron and hydrogen, a property exploited by Stock for analytical purposes.

The total bond energies of boron hydride molecules are indicated by their heats of atomization ΔH_a; the values in Table 4–1 were calculated assuming a heat of atomization for boron of 563 kJ/g atom and a heat of atomization for hydrogen of 218 kJ/g atom. The heats of atomization of the hydrides have been used to calculate bond energy terms for two- and three-centre boron–hydrogen and boron–boron bonds. The values given in Table 4–2 were derived as follows. The relative constancy of terminal boron–hydrogen bond lengths from one compound to another indicates that little variation is expected in the bond energy term for these, and a figure in the region of 375 kJ/mole is consistent both with thermodynamic data for a number of compounds containing terminal B–H bonds and also with the Pauling electronegativity-bond energy relationship. Using this value, and a value of 150 kJ/mole for the enthalpy of dissociation of diborane into two BH_3 fragments, one obtains a bond energy term for each BHB three-centre bond in diborane of 450 kJ/mole, $(375 + \tfrac{1}{2} \times 150)$. The consistency of these figures with the heat of atomization of diborane is shown by summing the six bond energy terms of the molecule (four terminal B–H bonds, two BHB bridge bonds):

$$\text{for } B_2H_6: \quad \Sigma E(B–X) = 4 \times E(B–H) + 2 \times E(BHB)$$
$$= 4 \times 375 + 2 \times 450 = 2400 \text{ kJ/mole} = \Delta H_a$$

Extending the argument, retaining the values 375 and 450 kJ/mole for E(B–H) and E(BHB) respectively, one obtains a bond energy term E(B–B) of 310 kJ/mole for the two-centre boron–boron bond of tetra-borane(10) (see Fig. 3–3 for the bonding representation):

$$\text{for } B_4H_{10}: \quad \Sigma E(B–X) = 6 \times E(B–H) + 4 \times E(BHB) + E(B–B)$$
$$= 6 \times 375 + 4 \times 450 + E(B–B)$$
$$= 4360 \text{ kJ/mole } (\Delta H_a) \text{ if } E(B–B) = 310 \text{ kJ/mole}$$

Similarly, the value of ΔH_a for pentaborane(11), in which there are two three-centre BBB bonds (see Fig. 3–5) requires the allocation of 380 kJ/mole to each of these bonds:

for B_5H_{11}: $\Sigma E(B-X) = 8 \times E(B-H) + 3 \times E(BHB) + 2 \times E(BBB)$

$\qquad\qquad\quad = 8 \times 375 + 3 \times 450 + 2E(BBB)$

$\qquad\qquad\quad = 5110 \text{ kJ/mole } (\Delta H_a) \text{ if } E(BBB) = 380 \text{ kJ/mole}$

The bond energy terms so derived (375, 450, 310, and 380 kJ/mole for B–H, BHB, B–B, and BBB bonds respectively) have been used to calculate $\Sigma E(B-X)$ for the other hydrides in Table 4–1 (B_5H_9, B_6H_{10}, and $B_{10}H_{14}$; see Figs 3–14, 3–8, and 3–10 respectively). Agreement between $\Sigma E(B-X)$ and ΔH_a is good for pentaborane(9) and hexaborane(10), but poor for decaborane(14). For all three compounds, ΔH_a exceeds $\Sigma E(B-X)$, which suggests that all three benefit from resonance stabilization, the extent of which may reasonably be expected to increase with the number of boron atoms in the molecular skeleton. A further source of discrepancies which has been ignored in this treatment is the variation in boron–boron distances from compound to compound, with which $E(B-B)$ and $E(BBB)$ may be expected to vary.

The various bond energy terms in Table 4–2 show that reactions in which boron–boron and boron–hydrogen bonds are replaced by bonds to such elements as nitrogen, oxygen, fluorine, and chlorine are likely to be exothermic. For example, oxidation of the boron hydrides to boric oxide and steam liberates considerable amounts of energy (ΔH_c in Table 4–1) some 70 kJ/g if solid boric oxide is produced, about 55 kJ/g if the oxide is gaseous; cf. ethane, which when oxidized to carbon dioxide and steam affords only 40 kJ/g. In practice, towards oxygen and moisture the hydrides range in sensitivity from spontaneously explosive reactivity in the case of diborane to comparative unreactivity in the case of decaborane (14). Several are decomposed by or react with stopcock grease. The more reactive volatile boranes clearly are substances which can be handled only in apparatus from which oxygen and moisture are rigorously excluded, and in which they are exposed as little as possible to greased joints or stopcocks. Though routine operations can be carried out using solutions under inert atmospheres in conventional apparatus, the need to separate complex mixtures of unknown products available sometimes only in milligramme quantities prompted Stock[2] to devise and develop techniques for their vacuum manipulation, techniques which find widespread application today in the handling of volatile substances, whether reactive or not.

4–2   Some vacuum manipulative techniques

The techniques Stock devised allowed substances to be purified by vacuum distillation, characterized by vapour density and pressure measurement, stored, made up into reaction mixtures and so on in suitably designed apparatus, made mainly of glass, from which the air was pumped to a pressure of 0·01 mm Hg or below. In a vacuum line, transference of a volatile substance is effected by cooling (using, for example, externally applied liquid air) that section of the apparatus to which the sample is to be moved; the sample diffuses to and condenses in the cold section. Greased stopcocks or other cut-off devices are needed to isolate sections of the apparatus so that condensed substances do not simply diffuse back when the coolant is removed. When he found that his hydride mixtures reacted with stopcock grease, Stock devised a valve in which only glass and mercury surfaces came into contact with the contents of the vacuum line. The workings of this valve, the mercury float valve, are illustrated in Fig. 4–1. When it is closed, the glass floats F fit snugly

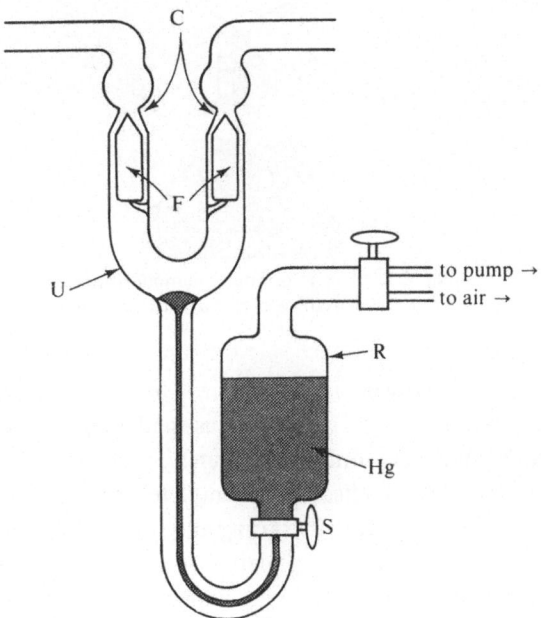

Fig. 4–1 A mercury float-valve

into ground constrictions C supported by the mercury Hg which, because of the snug fit of F into C, cannot leak further into the evacuated apparatus. The valve is opened by lowering the air pressure on the mercury reservoir R and opening stocock S until the mercury surface falls below the bottom of the U-tube U. Soft iron cores in the floats—a later modification— allow them to be moved from outside by means of a magnet if they happen to stick in the closed position.

A similar float, mounted upside-down, can be used as a one-way valve for noncondensable gases. One plays such a role in the apparatus shown

↑ to vacuum line

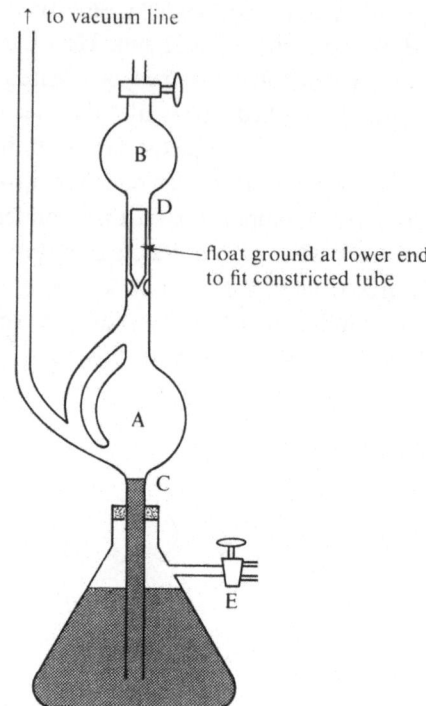

Fig. 4-2 A form of Töpler pump for transferring
a non-condensable gas into a measuring bulb

in Fig. 4–2, one of many devices which have been designed for the measurement in the vacuum line of a gas such as methane, nitrogen, or hydrogen which can be condensed only incompletely, if at all, by liquid air. Such a gas can be transferred effectively completely from section A into the measuring bulb B by repeatedly raising and lowering the mercury between levels C and D by alternate application of air pressure and suction to E.

Devices for opening, by breaking under vacuum, the capillary tip of a reaction vessel or storage tube or a thin-walled bulb, are shown in Fig. 4–3. Breakage is effected in both cases by rotating the handle H to which the breaker arm is attached. Alternatively, a small fragile glass bulb containing a sample can be sealed inside a larger vessel and then broken by a magnetic hammer—a glass-encased iron rod—raised from outside by a magnet which is subsequently moved away.

The low melting points of small samples can be measured using the apparatus shown in Fig. 4–4. The thin glass rod R incorporating a soft iron core C at its centre is raised by a magnet, a solid plug of sample is frozen in the tube at S, and the glass rod is lowered to rest on this plug of sample. The apparatus is allowed to warm slowly in a suitable cold bath in a Dewar vessel. Although the sample is unseen, its melting is registered by

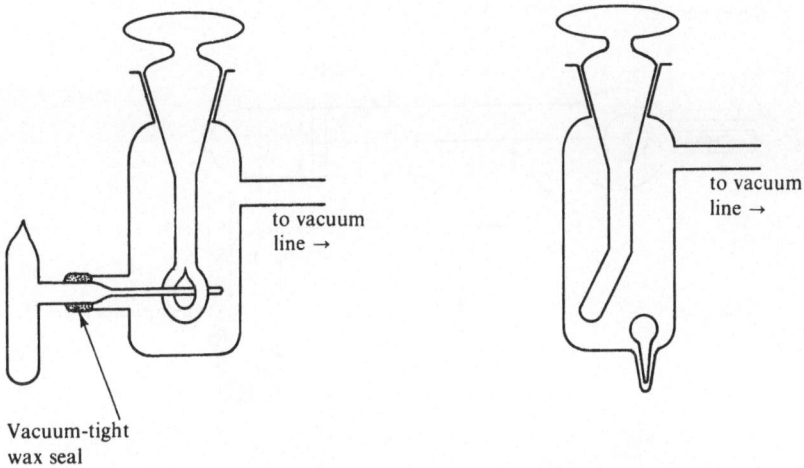

Fig. 4–3   Devices for breaking open glass vessels under vacuum

Vacuum-tight
wax seal

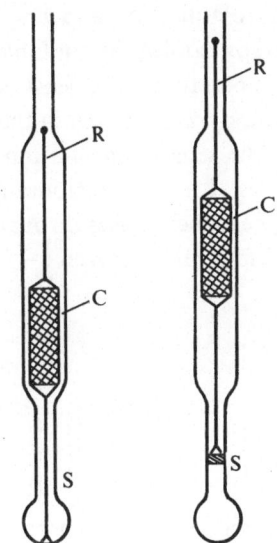

Fig. 4–4   Melting-point apparatus

the sinking of the glass rod R, the top end of which remains visible above the rim of the Dewar vessel.

Stock measured vapour densities of 0·05 mmole samples using a floating balance (Fig. 4–5) as a null instrument, adjusting the gas pressure to that needed to give sufficient buoyancy to the bulb B to balance the counterweight C as indicated by the pointer P; the balance arm pivots about needle points N. The gas pressure and temperature are read on manometer M and thermometer T.

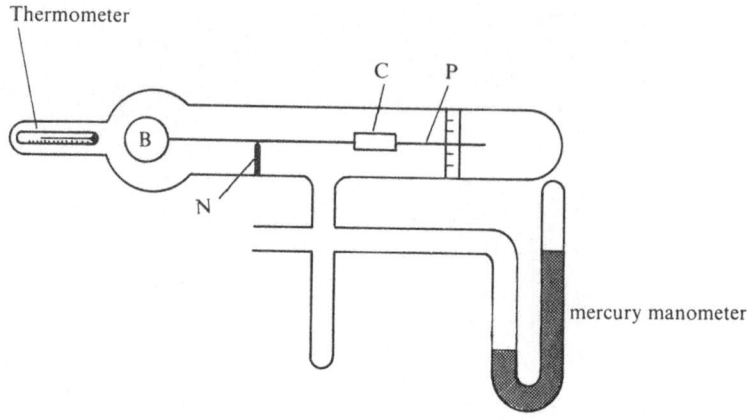

Fig. 4-5   Floating balance

Gas mixtures can be separated into their components (provided that these differ sufficiently in volatility) by allowing them to diffuse slowly through a series of U-traps cooled by suitable freezing mixtures to successively decreasing temperatures. The least volatile component of a mixture is collected in the first trap, the next in the next trap and so on as illustrated in Fig. 4-6 for the separation of a mixture containing penta-borane(11) (which condenses at $-78°$), tetraborane (which collects in the trap at $-126°$) and diborane, which passes through the first two traps but collects in the trap cooled in liquid nitrogen ($-196°$).

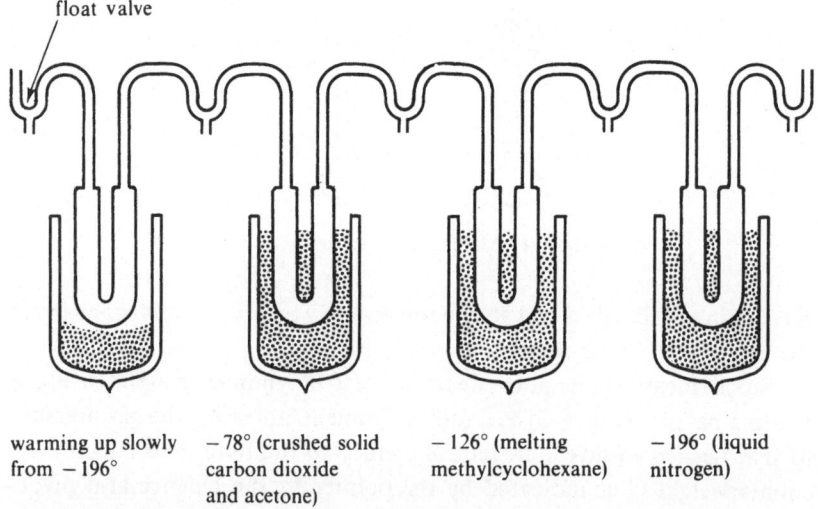

warming up slowly $-78°$ (crushed solid $-126°$ (melting $-196°$ (liquid
from $-196°$ carbon dioxide   methylcyclohexane)   nitrogen)
                                     and acetone)

Fig. 4-6   Trap-to-trap distillation to separate a mixture of diborane, tetraborane(10), and pentaborane(11)

It was with the aid of ingenious techniques such as these that Stock and his group made most impressive progress in their studies on the hydrides of boron and silicon, as described in Stock's book published in 1933.[2] Their ideas have been taken up and developed by many other workers,[3] but many are still extremely useful in their original or slightly modified form. However, by no means all volatile substances are unreactive towards mercury and so cannot be handled in apparatus equipped with float valves or mercury manometers. New materials have been developed which allow alternative solutions to such problems as are posed by reactivity to mercury or lubricants. For example, greased stopcocks can be replaced by unlubricated needle valves in which samples come into contact only with glass and fluorocarbon polymer surfaces. Nevertheless, Stock's influence will long be felt in this area of experimental chemistry.

4–3   Preparation and interconversions of the boron hydrides

The method originally used by Stock to prepare a mixture of boron and silicon hydrides was to treat magnesium boride (which normally contained some silicide) with hydrochloric acid. Using the sort of apparatus described above, he separated the condensable materials from the large volume of hydrogen also formed, and then isolated tetraborane, B_4H_{10}, as a major component of the condensable fraction. This when heated decomposed into a complex mixture of hydrides. Further studies on this mixture, and on the original gaseous product of hydrolysis of magnesium boride, resulted in the isolation and characterization of all the hydrides in Table 4–1. These same hydrides, particularly diborane, pentaborane(9), and decaborane(14), remain those about which most is known today. Though the hydrolysis of magnesium boride remains a practicable route to hexaborane(10), diborane is now the material from which most of the higher hydrides are prepared.

The most convenient laboratory preparations of diborane derive from the work of H. I. Schlesinger.[4] They make use of reactions between a hydride (for example NaH, $LiAlH_4$, or $NaBH_4$), and an acid, which must contain boron if the hydride does not (for example BF_3, BCl_3, $B(OMe)_3$, $AlCl_3$, $HgCl_2$, even HCl, H_2SO_4, or H_3PO_4). The reaction is normally carried out in an etheral solvent, preferably one in which the reactants are at least moderately soluble. Thus one convenient preparation involves the action of lithium aluminium hydride on an excess of boron trifluoride-diethyl ether complex in diethyl ether as solvent:

$$3LiAlH_4 + 4BF_3, OEt_2 \rightarrow 2B_2H_6 + 3LiAlF_4$$

although if the diborane is required free from ether the vacuum-line separation of the two is somewhat tedious, requiring a trap at $-100°$ or below to stop the ether, and the product contains a little ethyl fluoride as contaminant. The use of a high boiling ether avoids this problem, and for example the reaction between sodium tetrahydroborate and a Lewis acid in diglyme, $(MeOCH_2CH_2)_2O$, affords an easily purified product, for example,

$$3NaBH_4 + 4BF_3 \rightarrow 3NaBF_4 + 2B_2H_6$$

The best procedure is to add the tetrahydroborate solution *to* the boron trifluoride solution. The reverse procedure results in the formation of the heptahydrodiborate anion $B_2H_7^-$ and therefore a low yield of diborane. The use of a protic acid in a reaction requiring no solvent is illustrated by the following equation:

$$2NaBH_4 + H_2SO_4 \rightarrow B_2H_6 + 2H_2 + Na_2SO_4$$

Care is needed in all these exothermic reactions to avoid too rapid mixing of the reactants which can lead to dangerously violent reaction.

The boron trihalide–alkali metal tetrahydroborate reaction has also been used for the industrial scale preparation of diborane. Large-scale routes to diborane and ultimately alkylated higher boranes were particularly vigorously explored in the United States during the nineteen-fifties when it appeared worth making such substances as high energy fuels.[5] In principle, the combustion of boranes or alkylboranes is capable of affording some 30 kJ more energy per gramme than hydrocarbon fuels (see Table 4–1). In practice, however, it is difficult to effect complete combustion in a suitable motor, so not all of this energy is released. Oxidation tends to stop at the composition BO, an oligomeric or polymeric substance of so low volatility as to tend to block exhausts. Nevertheless although boron hydrides are not being prepared industrially in the quantities that once appeared possible, the feasibility of preparing diborane on a large scale by the following other routes has been demonstrated.

$$B(OR)_3 \longrightarrow HB(OR)_2 \xrightarrow{\text{heat}} B_2H_6 + B(OR)_3$$

$$BCl_3 \xrightarrow[\text{Ag catalyst}]{H_2; 700°} B_2H_5Cl + HBCl_2 \xrightarrow[\text{disproportionation}]{\text{thermal}} B_2H_6 + BCl_3$$

This last reaction, the hydrogenation of boron trichloride, is an adaptation of another of Schlesinger's methods.

Diborane, prepared by one or other of the above routes, can be converted into higher boranes by moderate heat. The thermal decomposition of diborane with or without added hydrogen leads to complex mixtures of higher hydrides and hydrogen, but yields of a particular hydride can be optimized by suitable choice of temperature, pressure, hydrogen pressure, and reaction time. Much effort has been devoted to sorting out the reactions that occur. These are many and varied and somewhat akin to the thermal cracking of hydrocarbons, except that they take place at lower temperatures and apparently involve a number of key intermediates which have never been isolated, though their existence has been inferred from kinetic studies, indicated by mass spectroscopic studies, and is more or less consistent with Lipscomb's topological treatment of boron hydrides. Though there is not yet general agreement about several aspects of the interconversion reactions suggested by different groups of workers, a few common features of their findings have emerged.

Reference to Table 3–2 shows that all the known boron hydrides are *even electron* molecules. That is, they have formulae B_pH_{p+q} in which q is invariably *even* for the neutral molecules. In the absence of evidence to the contrary, it is generally assumed that the thermal decomposition of these hydrides involves only even-electron species as intermediates, not free radicals. Whether or not the intermediates are isolable hydrides, they all can be expressed by the formula B_pH_{p+q}, with q even. This being so, it follows that interconversion or exchange reactions of such species must occur by the addition, elimination, or transfer of fragments which themselves conform to the same rule (B_pH_{p+q}; q even). Thus when hydrogen alone unaccompanied by boron ($p = 0$) is added to or removed from a molecule, it must be in the form of molecular hydrogen ($q = 2$). When $p = 1$, q can equal one (BH eliminated or added) or three (BH_3 eliminated or added). Indeed the four main types of reaction of boranes at moderate temperatures (below 200°C) are believed to be as follows:

    (i) the exchange of BH groups;
    (ii) the exchange of BH_3 groups;
    (iii) the elimination of BH_3 molecules by unimolecular dissociation;
    (iv) condensation with elimination of H_2.

These points are illustrated by the series of interconversion reactions believed to be involved in the formation of B_3, B_4, and B_5 species from diborane, for which the following reaction scheme (due to L. H. Long) has been suggested:[6]

$$2B_2H_6 \rightleftharpoons BH_3 + B_3H_9 \qquad (4\text{–}1)$$

$$BH_3 + B_2H_6 \rightleftharpoons B_3H_9 \qquad (4\text{–}2)$$

$$B_3H_9 \rightleftharpoons B_3H_7 + H_2 \qquad (4\text{–}3)$$

$$2B_3H_9 \rightarrow 3B_2H_6 \qquad (4\text{–}3a)$$

$$BH_3 + B_3H_7 \rightleftharpoons B_4H_{10} \qquad (4\text{–}4)$$

$$B_2H_6 + B_3H_7 \rightarrow BH_3 + B_4H_{10} \rightleftharpoons B_5H_{11} + H_2 \qquad (4\text{–}5, 8)$$

$$B_3H_9 + B_3H_7 \rightarrow B_2H_6 + B_4H_{10} \qquad (4\text{–}6)$$

$$2B_3H_7 \rightarrow B_2H_6 + B_4H_8 \tag{4-7}$$

$$B_3H_9 + B_4H_{10} \rightarrow B_2H_6 + B_5H_{11} + H_2 \tag{4-9}$$

$$B_3H_7 + B_4H_{10} \rightarrow B_2H_6 + B_5H_{11} \tag{4-10}$$

$$H_2 + B_4H_8 \rightarrow B_4H_{10} \tag{4-11}$$

$$BH_3 + B_4H_8 \rightleftharpoons B_5H_{11} \tag{4-12}$$

$$B_3H_9 + B_4H_8 \rightarrow B_2H_6 + B_5H_{11} \tag{4-13}$$

Similar reactions have been suggested to account for the formation of the higher hydrides. The reaction pattern is more readily discerned from the diagrammatic representation shown in Fig. 4–7. Note that of the species

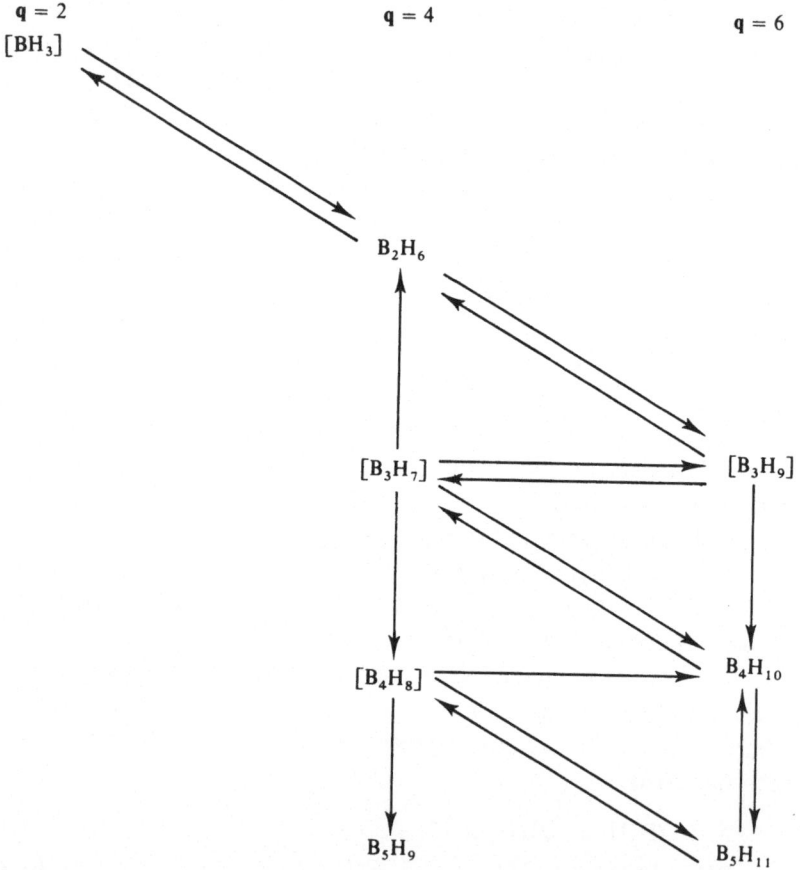

Horizontal lines represent addition or elimination of H_2 [see text, equations (4–3), (4–8), (4–9), and (4–11)].

Vertical lines represent addition or abstraction of [BH] [see text, equations (4–6)–(4–10)].

Slanting lines represent addition or removal of [BH_3] [see text, equations (4–1), (4–2), (4–3a), (4–4), (4–5), (4–6), (4–9), (4–12), and (4–13)].

Fig. 4–7 Likely reactions in the thermal decomposition of diborane, involving various species B_pH_{p+q}

involved in these reactions, whereas B_2H_6, B_4H_{10}, B_5H_9, and B_5H_{11} are known compounds, the species BH_3, B_3H_7, B_3H_9, and B_4H_8 are apparently too susceptible to such reactions to be isolated and characterized, although BH_3, B_3H_7, and B_4H_8 fragments are all known in the form of adducts with Lewis bases. Possible topologies for B_3H_7, B_3H_9, and B_4H_8 were shown in Fig. 3–12.

The reaction scheme illustrated differs in one important respect from earlier schemes, which postulated the unimolecular dissociation of diborane into two borane fragments as the first step, a process which now appears likely to involve a higher activation energy than was originally thought. The enthalpy change for the dissociation $B_2H_6 \rightarrow 2BH_3$ is now known to be about 150 kJ—earlier work had indicated a figure nearer 115 kJ—and the activation energy for the dissociation can be expected to be greater still. However, whatever the precise mode of its formation, monoborane(3), BH_3, is believed to be one of the first reactive intermediates in the thermal decomposition of diborane. The rate-controlling step is believed to be reaction (4–3), the relatively slow reversible elimination of hydrogen from B_3H_9. This is consistent with the observation that the disappearance of diborane obeys a 3/2-order rate law, since summation of reactions (4–1) and (4–2) shows that two B_3H_9 molecules are reversibly generated from three B_2H_6 molecules.

Because the thermal decomposition of diborane involves such a complicated set of inter-related equilibria, the identity and proportions of the products which can ultimately be isolated depend critically on the initial partial pressures of hydrogen and diborane, the temperature, and the reaction time. Pentaborane(11) is a major product in the early stages of the reaction. Pentaborane(9) and ultimately decaborane(14) are later major constituents. The conditions suited to the formation of the commoner boranes are as follows:

Tetraborane(10), B_4H_{10}; diborane undergoes 20% conversion into 90% B_4H_{10} during about five hours at 200 atm at 80–90°C.

Pentaborane(9), B_5H_9; passage of diborane in a four-fold excess of hydrogen through reactors at 200–240° requires a residence time of only a few seconds to achieve good yields of pentaborane(9).

Pentaborane(11), B_5H_{11}, can be prepared by passing diborane in hydrogen at 2 atmospheres during two minutes through a reactor at 100–120°. Samples relatively free from other higher hydrides result from heating pure diborane at 150–160° and 1 atmosphere for 2–10 seconds, though the percentage conversion is low.

Hexaborane(10), B_6H_{10} is produced in only very low yield and contaminated by other products from the pyrolysis of diborane, but together with tetraborane(10) is a major product (6% yield) of the phosphoric acid hydrolysis of magnesium boride (Mg_3B_2 or MgB_2).

Decaborane(14), $B_{10}H_{14}$, like pentaborane(9), is conveniently prepared by hot tube (160–200°C) pyrolysis of diborane, though longer heating is required to ensure the accumulation of the higher borane intermediates needed for good yields, and high concentrations of hydrogen are best avoided as these favour formation of pentaborane(9) and reduce the reaction rate.

In all cases the products can be separated and purified by gas chromatography or by trap-to-trap distillation in the vacuum line using the trap temperatures shown in Table 4–3.

Table 4–3 **Trap temperatures suitable for the separation of boron hydrides in the vacuum line**

Hydride	t_1 °C	t_2 °C	Cooling bath for t_2
B_2H_6	−126	−196	liquid nitrogen
B_4H_{10}	−95	−126	melting methylcyclohexane
B_5H_9	−78	−95	melting toluene
B_5H_{11}	−63	−78	powdered carbon dioxide in acetone
B_6H_{10}	−35	−63	melting chloroform
$B_{10}H_{16}$	−20	−35	melting 1,2 dichloroethane
$B_{10}H_{14}$	—	−20	melting carbon tetrachloride

t_1 °C = trap temperature suitable for separating a particular hydride from less volatile substances; the hydride in question passes through
t_2 °C = trap temperature suitable for separating a particular hydride from more volatile substances; the hydride in question condenses

4–4 Some reactions of diborane

Diborane is a highly reactive substance. It is liable to inflame or even explode when exposed to air, the oxygen of which oxidizes it to boric oxide and steam. At low concentration in air, it has a characteristic, rather unpleasant smell, and is toxic. It is violently hydrolysed to hydrogen and boric acid by water, reacts instantaneously with a wide range of Lewis bases, and is a powerful reducing agent. Its smooth addition to olefins in ether solvents, the hydroboration reaction,[13] makes it a valuable reagent in organic syntheses, since the organoboranes produced can subsequently be converted into a wide range of functionally substituted products. It is itself an important intermediate in the preparation of many other boron–hydrogen compounds.

In this section, its reactions are grouped as follows:

1. *Formation of adducts with donor molecules L:*
Diborane can act as a Lewis acid, suffering cleavage of one BHB bridge bond for every coordinate link formed:

$$B_2H_6 \underset{-L}{\overset{+L}{\rightleftharpoons}} L(H_2)BHBH_3 \underset{-L}{\overset{+L}{\rightleftharpoons}} 2L,BH_3 \quad \text{or} \quad L_2BH_2^+BH_4^-$$

2. *Reactions with compounds* $H(\delta+)-X(\delta-)$ *to eliminate hydrogen*:
Compounds $H(\delta+)-X(\delta-)$ which can in principle act as protic acids
react with diborane to eliminate hydrogen and attach the conjugate base
X^- to boron:

$$>B(\delta+)-H(\delta-)+H(\delta+)-X(\delta-) \rightarrow >B-X+H_2$$

3. *Addition of B–H to unsaturated groups—hydroboration*:
The B–H groups of diborane add across multiple bonds with an orientation that reflects their polarity:

$$>B(\delta+)-H(\delta-)+Y(\delta+) = Z(\delta-) \rightarrow >B-Z-Y-H$$

4. *Exchange reactions*:
Diborane exchanges hydrogen for another atom or group in reactions
with molecular hydrogen, other boron compounds, or suitable derivatives
of other elements:

$$>B-H+M-X \rightarrow >B-X+M-H$$

5. *Oxidative cleavage of B–H bonds*:
The B–H bonds of diborane are cleaved by oxygen or halogens:

$$>B-H+X_2 \rightarrow >B-X+HX$$

6. *Reduction of diborane*:
The addition of electrons to diborane leads to a disproportionation
reaction:

$$B_2H_6+2e^- \longrightarrow [B_2H_6^{2-}] \xrightarrow{B_2H_6} BH_4^- + B_3H_8^-$$

7. *Decomposition to hydrogen and higher hydrides*:
When heated or otherwise supplied with energy, diborane generates
hydrogen and various higher hydrides (see section 4–3). If this occurs in
the presence of another reagent, for example an acetylene, products of
reaction with species containing several boron atoms may result (see
Chapter 6)

$$B_2H_6+RC\vdots CR \xrightarrow{\text{high energy source}} H_2+\text{alkylboranes}+\text{carboranes}$$

Examples of these various types of reaction are given in Table 4–4.

Table 4–4 Some reactions of diborane

1 Formation of adducts with donor molecules

$B_2H_6+L \rightarrow L(H_2)BHBH_3$ (L = Me_3N, H^-, and probably many other Lewis bases)

$B_2H_6+X^- \rightarrow [H_3BXBH_3]^-$ (X = CN^-, Me_2N^-, H_2P^-, HS^-)

$B_2H_6+2L \xrightarrow[\text{cleavage}]{\text{symmetrical}} 2L,BH_3$ (L = R_3N, R_3P, R_3As, R_2O, R_2S, CO, PF_3, H^-, F^-, OR^-)

$B_2H_6+2L \xrightarrow[\text{cleavage}]{\text{unsymmetrical}} L_2BH_2^+BH_4^-$ (L = NH_3, $MeNH_2$, Me_2NH, Me_2SO, H_2O?)

Table 4-4 (cont.)

2 Reactions with compounds $H(\delta+)-X(\delta-)$ to eliminate hydrogen

$$B_2H_6 + HX \rightarrow H_2BH_2BHX + H_2 \quad (X = Cl, Br)$$

$$B_2H_6 + HX \rightarrow H_2B\underset{X}{\overset{H}{\diagdown\diagup}}BH_2 + H_2 \quad (X = NR_2, PR_2, OR, SR)$$

$$B_2H_6 + 2HX \rightarrow 1/n(H_2BX)_n + 2H_2 \quad (X = NR_2, PR_2)$$

$$B_2H_6 + 2NH_3 \rightarrow (H_2BNH_2)_n, (HBNH)_n, \text{ and } (BN)_n$$

$$B_2H_6 + 4MeOH \rightarrow 2HB(OMe)_2 + 4H_2$$

$$B_2H_6 + \text{excess } HX \rightarrow 2BX_3 + 6H_2 \quad (X = OH, OR, Cl, Br)$$

3 Addition of B–H to unsaturated groups—hydroboration

$$B_2H_6 + RCH:CH_2 \xrightarrow{\text{ethers}} (RCH_2CH_2)_nB_2H_{6-n} \quad (n = 1 \rightarrow 4) + (RCH_2CH_2)_3B$$

$$B_2H_6 + RC:CR \xrightarrow{\text{ethers}} (RCH:CR)_nB_2H_{6-n}$$

$$B_2H_6 + R^1R^2C:O \longrightarrow (R^1R^2CHO)_3B \quad (R^1, R^2 = H, \text{alkyl, aryl, OR, Cl})$$

$$B_2H_6 + RC:N \longrightarrow RC:N,BH_3 \longrightarrow (RCH_2NBH)_3$$

$$B_2H_6 + CO \xrightarrow{\text{NaBH}_4} (CH_3BO)_3$$

4 Exchange reactions

$$B_2{}^1H_6 + B_2{}^2H_6 \rightarrow B_2{}^1H_n{}^2H_{6-n} \quad (n = 1 \rightarrow 5)$$

$$B_2{}^1H_6 + {}^2H_2 \rightarrow B_2{}^1H_5{}^2H + {}^1H^2H$$

$$B_2H_6 + Me_3B \rightarrow B_2H_nMe_{6-n} \quad (n = 2 \rightarrow 5; \text{Me terminal only})$$

$$B_2H_6 + \text{limited } BX_3 \rightarrow H_2BH_2BHX \quad (X = Cl, Br)$$

$$B_2H_6 + \text{excess } BX_3 \rightarrow HBX_2 \quad (X = F, Cl, Br)$$

$$B_2H_6 + Me_6Al_2 \rightarrow Al(BH_4)_3 + Me_3B$$

$$B_2H_6 + LiOR \rightarrow LiBH_4 + (RO)_3B$$

5 Oxidative cleavage of B–H bonds

$$B_2H_6 + \text{excess } O_2 \rightarrow B_2O_3 + 3H_2O$$

$$B_2H_6 + \text{limited } X_2 \rightarrow B_2H_5X + HX \quad (X = Cl, Br)$$

$$B_2H_6 + \text{excess } X_2 \rightarrow 2BX_3 + 6HX$$

6 Reduction of diborane

$$B_2H_6 + Na/Hg \rightarrow [B_2H_6^{2-}](?) \rightarrow NaBH_4 + NaB_3H_8$$

7 Decomposition to hydrogen and higher hydrides

$$B_2H_6 \xrightarrow[\text{electric discharge}]{\text{heat, u.v. or}} H_2 + \text{higher hydrides}$$

$$B_2H_6 + RC:CR \xrightarrow[\text{electric discharge}]{\text{heat, u.v. or}} H_2 + \text{organoboranes} + \text{small carboranes}$$

1   Formation of adducts with donor molecules

The relatively strong Lewis acidity of diborane is an important source of its reactivity. Many of its reactions which do not end simply in adduct formation nevertheless involve coordination complexes as intermediates. The high susceptibility of diborane to nucleophilic attack is not unexpected in a substance that is relatively deficient in electrons, but it is significant that molecular rearrangement accompanies adduct formation. Diborane as a molecule does not have a vacant molecular orbital in

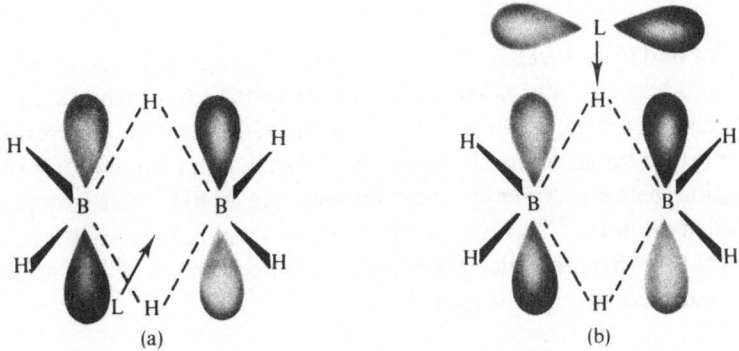

                        (a)   (b)

Fig. 4–8   Diagrams showing how the vacant π^* $1b_{2g}$ orbital of diborane has
(a) the wrong symmetry for a bonding interaction with a σ- or π-donor approaching along the two-fold axis perpendicular to the BH_2B bridge plane, but
(b) appropriate symmetry for a bonding interaction with a π-donor approaching along the two-fold axis through the bridging hydrogens

which it can accommodate an extra pair of electrons to form a stable adduct L,B_2H_6 in which the structure $H_2BH_2BH_2$ is retained. Reference to Figs 2–16 and 2–18 shows that the lowest energy unfilled orbital of diborane is the $1b_{2g}$ orbital that results from the π^* combination of the boron p_x orbitals; this orbital has the wrong symmetry to interact either with the type of σ-donor with which diborane forms a variety of complexes (see Table 4–4), or with a π-donor located on the two-fold axis perpendicular to the plane of the BH_2B bridge, as in Fig. 4–8a. Its symmetry *is* appropriate for a bonding interaction with a π-donor approaching along the two-fold axis through the bridging hydrogen atoms (Fig. 4–8b), and approach of π-donors along this axis may well provide a mechanism for hydroboration reactions. The formation of a stable adduct with the structure shown in Fig. 4–8b is, however, unlikely in view of the repulsive effect of the upper bridging hydrogen atom. The complexes L,B_2H_6 through which many of the reactions of diborane are believed to proceed apparently have a σ-donor, L, coordinated to one of the two boron atoms in the molecule, which remains linked to the other boron by only one

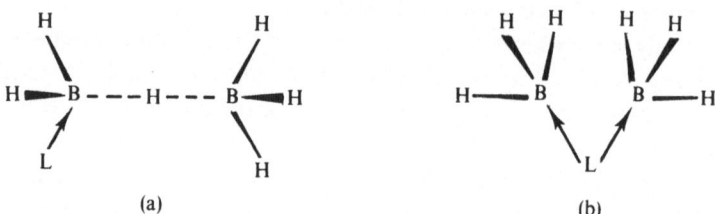

Fig. 4-9 Structure of complexes L, B_2H_6:
(a) The ligand atom of L uses one pair of electrons to form one σ-dative link
(b) The ligand atom of L uses two pairs of electrons to form two σ-dative links

BHB bond (Fig. 4–9a). Although the molecule rearranges too rapidly for the two boron atoms to be readily distinguished from each other in the ^{11}B n.m.r. spectrum unless such a system is held at a low temperature, support for the structure shown in Fig. 4–9a has come from a study of the reaction between trimethylamine-borane, Me_3N,BH_3, and isotopically labelled diborane $^{10}B_2H_6$ at $-78°$. These reversibly generate the adduct $Me_3N,^{11}B^{10}BH_6$ in which the unlabelled boron remains firmly attached to the trimethylamine nitrogen :[14]

$$Me_3N,^{11}BH_3 + 0{\cdot}5\ ^{10}B_2H_6 \underset{}{\overset{-78°}{\rightleftharpoons}} \underset{\underset{NMe_3}{\uparrow}}{H_2B\cdots H\cdots{}^{10}BH_3} \nrightarrow Me_3N,^{10}BH_3 +$$

$$0{\cdot}5\ ^{11}B_2H_6$$

An interesting example of a specific adduct L,B_2H_6 is the anion $B_2H_7^-$ (L = H^-), which results from reactions between diborane and alkali metal hydrides or tetrahydroborates:

$$MH \xrightarrow{\ 0{\cdot}5\ B_2H_6\ } MBH_4 \xrightarrow{\ 0{\cdot}5\ B_2H_6\ } MB_2H_7$$

Its formation is the cause of the low yield when the preparation of diborane is attempted by the addition of boron trihalide to an ethereal solution of alkali metal tetrahydroborate, instead of vice versa. Its single hydrogen bridge, which is presumably linear, gives rise to an infrared absorption at $2050\ cm^{-1}$. Related anions $R_3BHBR_3^-$ result from reactions between lithium hydride and trialkylborane. For all such complexes, formally removing the bridging hydrogen as a proton leaves the residual anion $H_3BBH_3^{2-}$ or a derivative thereof, species isoelectronic, and so presumably isostructural, with ethane.

A different structure, shown in Fig. 4–9b, is possible for adducts L,B_2H_6 in which the Lewis base L has *two* pairs of electrons available for donation as have the anions Me_2N^-, H_2P^- and HS^-. The cyanide ion can also coordinate to two monoborane (3) units simultaneously, though in a linear complex $H_3BCNBH_3^-$.

Adducts of stoichiometry L,B_2H_6 normally represent only the first stage in the interaction of donor molecules with diborane, and can be isolated only when an excess of diborane is present. Use of an excess of Lewis base leads to adducts of stoichiometry L,BH_3. These can have two types of structure, either the covalent monoborane (3) adduct structure $L \rightarrow BH_3$, or the ionic boronium tetrahydroborate structure $L_2BH_2^+BH_4^-$, depending on whether the second mole of base coordinates to a different boron atom, or the same boron atom, as the first:

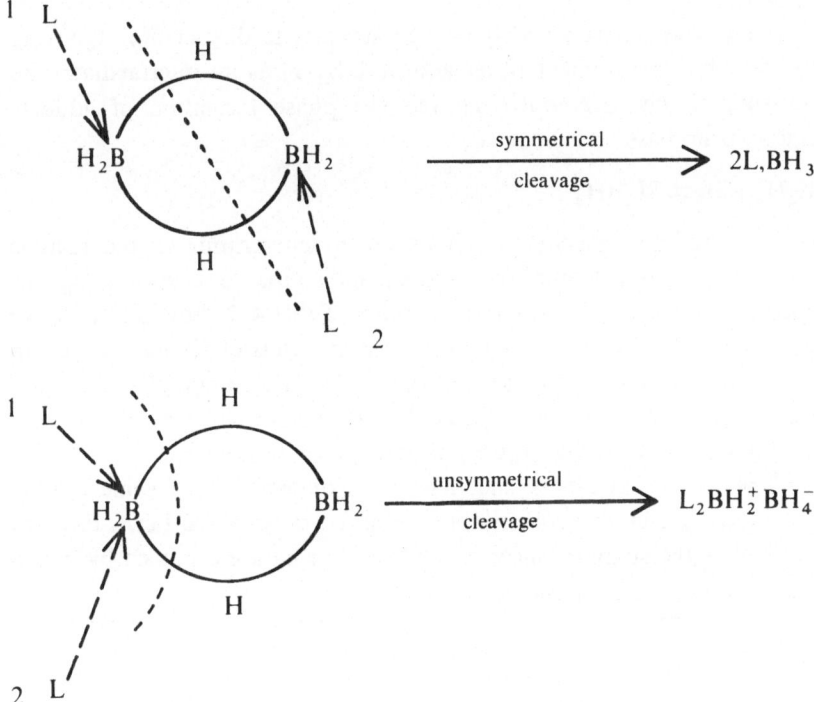

In a strongly coordinating solvent like tetrahydrofuran, diborane is believed to exist primarily in the form of the covalent adduct C_4H_8O,BH_3. The electrical conductivity of the solution is, however, consistent with the presence also of some of the ionic form $(C_4H_8O)_2BH_2^+BH_4^-$. The following equilibria probably operate:

In diethyl ether, in which its solubility barely exceeds that predicted by Raoult's law, diborane exists predominantly in the uncoordinated form B_2H_6. The respective positions of equilibrium depend on the donor strength of L, and also its bulk. If the Lewis base is small and more electronegative than hydrogen, the second molecule of base becomes attached to the same boron as the first. Thus, ammonia, methylamine, and dimethylamine all react with diborane at low temperature to form ionic adducts $L_2BH_2^+BH_4^-$. A larger base or one more electropositive than hydrogen attacks the second boron atom of diborane. Trimethylamine for example gives the covalent adduct L,BH_3, as do the majority of bases.

These adducts vary widely in their stability to dissociation. Ignoring the possible involvement of an adduct L,B_2H_6 as an intermediate, the pressure change accompanying the gas phase formation of adducts L,BH_3 from base and diborane,

$$B_2H_6 + 2L \rightleftharpoons 2L,BH_3$$

can be used to measure the change with temperature in the relative concentration of reactants and product, and so the free energy change for such reactions. The bond energy terms in Table 4–2 show that a dative link from a Lewis base to one of the boron atoms of diborane needs an energy of only some 75–80 kJ/mole for the cleavage of the BHB bridging link to be exothermic. Quite feeble donors accordingly generate measurable quantities of adduct L,BH_3 in reactions with diborane.

The vapour pressures and degrees of dissociation of some borane complexes are in Table 4–5.[15] Note that for most of the Lewis bases in Table 4–5, the spare valences of the ligand atom are occupied by bonds

Table 4–5 Some borane adducts L,BH₃

Compound	Vapour pressure	Dissociation in vapour	Compound	Vapour pressure	Dissociation in vapour
Me_3N,BH_3	1 mm at 27°	nil at 200°	Me_2O,BH_3	18 mm at −78°	Near 100%
Me_3P,BH_3	3 mm at 50°	slight >200°	Me_2S,BH_3	4 mm at −78°	60% at 53°
Me_3As,BH_3	7 mm at 47°	slight at 120°	OC,BH_3	320 mm at −78°	extensive
Me_3Sb,BH_3	—	complete	F_3P,BH_3	267 mm at −80°	extensive

to chemically unreactive methyl groups. If hydrogen atoms were attached to the ligand atom, irreversible loss of hydrogen (as from Me_2NH,BH_3 to form Me_2NBH_2) would render pressure measurements useless as a means of studying dissociation equilibria, and multiple bonds to ligand nitrogen or oxygen (as in $RC\vdots N$ or $R_2C\vdots O$) would make the Lewis base susceptible to hydroboration (cf. Table 4–4 sections 2 and 3). One notable feature of the data in Table 4–5 is that the second row elements phosphorus and sulphur form adducts virtually as stable as those of the first row elements nitrogen and oxygen, though there is a falling off at arsenic. This behaviour is reminiscent of transition metal complexes in which 'back coordination' from the acceptor to the donor may occur. That monoborane(3), BH_3, forms stable adducts with 'π-acid ligands' like trimethylphosphine and dimethylsulphide, and forms adducts at all with çarbon monoxide and phosphorus trifluoride (CO and PF_3 do not coordinate to boron halides) reflects its ability to function as a σ-acceptor and π-donor. This property is analogous to the hyperconjugative effect of a methyl group, and is illustrated by regarding borane–carbonyl as a resonance hybrid to which the following canonical forms contribute:

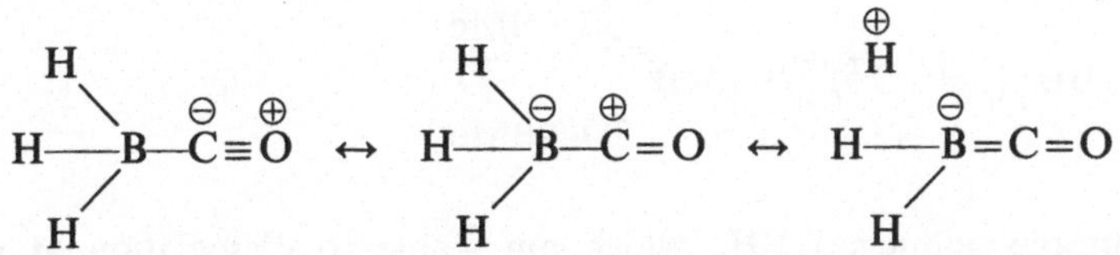

That the boron–carbon bond of borane–carbonyl has some multiple bond character is supported by recent SCF–MO calculations, and by its photo-electron spectrum.[16] The bond distances and angles of the molecule are B–H, 119 pm; B–C, 154 pm; C–O, 113 pm; and $\angle$ HBH, 114°, i.e. *greater* than the tetrahedral angle (by microwave spectroscopy; cf. B–C, 160 pm for boron–carbon single bonds in various alkylboron compounds, and C–O, 113 pm for uncoordinated carbon monoxide).

Borane–carbonyl is a member of an interesting family of borane complexes the structures of which underline simultaneously the iso-electronic relationship between BH_3 and CH_3^+, and between mono-borane(3) and the oxygen atom as members of a series of species formally capable of forming double bonds; $=O$, $=NH$, $=CH_2$, $=BH_3$. For example compare the linear skeletons of H_3B,CO, $(CH_3CO)^+$, and OCO. Borane–carbonyl even resembles carbon dioxide chemically to the extent of being absorbed by alkali with formation of salts of the anion $H_3BCO_2^{2-}$ (cf. $H_3CCO_2^-$ or CO_3^{2-}). These and other relationships may be summarized thus:

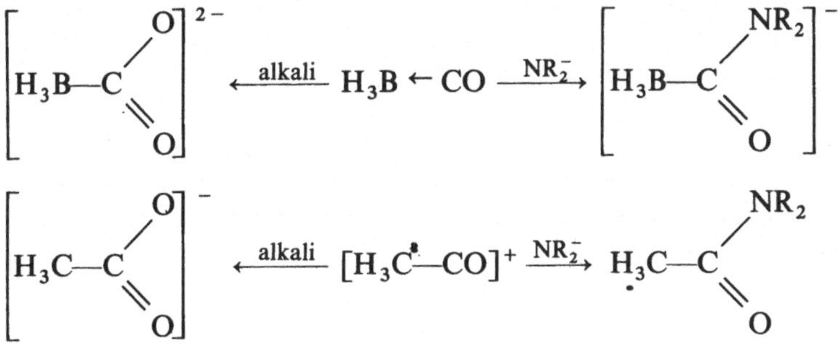

$$\left[\begin{array}{c} O-C \underset{O}{\overset{O}{\diagup}} \end{array} \right]^{2-} \xleftarrow{\text{alkali}} O=C=O \xrightarrow{\quad NR_2^-\quad} \left[\begin{array}{c} O-C \underset{O}{\overset{NR_2}{\diagup}} \end{array} \right]^{-}$$

A different type of reaction between diborane and carbon monoxide is that which occurs in tetrahydrofuran solution in the presence of sodium borohydride; this provides a convenient synthesis of trimethylboroxine, $(MeBO)_3$:

$$3[BH_3] + 3CO \xrightarrow[\text{THF}]{\text{NaBH}_4} MeB \underset{O-\overset{}{B}Me}{\overset{O-BMe}{\diagup\!\!\!\diagdown}} O$$

Borane adducts L, BH_3 which are stable to dissociation at room temperature are air- and moisture-stable substances, and so provide a convenient form in which to store diborane, which can be released as required by displacing borane(3) with a stronger Lewis acid, for example boron trifluoride. Alkali metal tetrahydroborates $M^+BH_4^-$ (effectively adducts in which $L = H^-$) and amine–boranes R_3N, BH_3 are commonly used examples. In such adducts, the boron atoms are four coordinate and involved in no three-centre bonds, and so not susceptible to nucleophilic attack (cf. the carbon atoms of isoelectronic alkanes, CH_4 or R_3CCH_3). Their boron-attached hydrogen atoms are much less susceptible to cleavage by weak protic acids than are the hydrogen atoms of diborane, but nevertheless are capable of interactions with electrophiles that arise from their hydridic, $B(\delta+)-H(\delta-)$, character. Their tendency to form adducts $L, 2BH_3$ ($M^+ H_3B\cdots H\cdots BH_3^-$ or $R_3N, B(H_2)\cdots H\cdots BH_3$) with an excess of diborane is one example of this. Another example of the interaction of coordinated borane with an electrophile appears to be that which causes a low frequency O–H stretching absorption to appear in the infrared spectra of alcohols when these are mixed with borane adducts in carbon tetrachloride solution.[17] A low O–H stretching frequency is also found when alcohols are mixed with species like amines, and the hydrogen of the hydroxyl group is involved in hydrogen bonding, i.e. both $L, BH_3 \cdots H-OR$ and $R_3N \cdots\cdots H-OR$ have low $\nu(O-H)$.

2   Reactions with compounds H(δ+)—X(δ−) to eliminate hydrogen

The reactivity of diborane towards reagents in which hydrogen is bound to a relatively electronegative element arises from the low energy and slight polarity $B(\delta+)-H(\delta-)$ of its boron–hydrogen bonds, which are readily replaced by bonds to the more electronegative element. The vigorous hydrolysis of diborane that proceeds spontaneously below room temperature is one such reaction:

$$B_2H_6 + 6H_2O \rightarrow 2B(OH)_3 + 6H_2$$

At very low temperatures, an adduct $B_2H_6,2H_2O$ can be prepared. If ice and a measured excess of diborane are condensed together at $-196°$ in the vacuum line, and the temperature is then raised to about $-130°$, unreacted diborane (but no hydrogen) can be pumped from the system, leaving a residue of composition $B_2H_6,2H_2O$. This may be either H_2O,BH_3 or $(H_2O)_2BH_2^+ BH_4^-$. It subsequently loses hydrogen only on further warming, though still at well below room temperature.[18]

Similar behaviour is found in the ammonia–diborane system, though in this case an appreciably higher temperature is needed before hydrogen is evolved. The adduct $(NH_3)_2BH_2^+ BH_4^-$ (the so-called 'di-ammoniate' of diborane, bis(ammonia)dihydroboronium tetrahydroborate), prepared by condensing ammonia and diborane together at low temperature, is a crystalline solid at room temperature and loses hydrogen only when heated above about 30°C.

$$B_2H_6 + 2NH_3 \xrightarrow{-96°} (NH_3)_2BH_2^+ BH_4^- \xrightarrow{90°} H_2 + \text{various B–N}$$
$$\text{compounds.}$$

The ease of cleavage of hydrogen from diborane appears to increase with the electronegativity of the atom to which the protic hydrogen is initially attached. Thus, bonds $H(\delta+)$—$X(\delta-)$ cleave $B(\delta+)$—$H(\delta-)$ bonds with increasing readiness in the sequences:

$$H–NR_2 < H–OR < H–F$$

$$H–I < H–Br < H–Cl < H–F$$

No adduct is isolable in the hydrogen chloride–diborane system, from which the following products (together with hydrogen) may be obtained by the use of progressively increasing proportions of hydrogen chloride; monochlorodiborane, B_2H_5Cl; dichloroborane, $HBCl_2$; and boron trichloride, BCl_3. It is interesting that the BH_2B bridge of diborane persists on halogenation only in the case of the mono-halogeno derivative B_2H_5X, in which the halogen atom occupies a terminal position. Dichloroborane $HBCl_2$ does not associate to form tetrachlorodiborane $Cl_2BH_2BCl_2$, nor do any other dihalogenoboranes HBX_2 or dialkoxyboranes $HB(OR)_2$ associate. Moreover, the only *di*halogeno*di*borane known is the di-iodo compound $IHBH_2BHI$. A likely explanation is that dative $B \leftrightharpoons X$ π-bonding from the halogen to the vacant boron p-orbital of planar monomeric dihalogenoboranes HBX_2 effectively stabilizes these monomers relative to the hypothetical dimer $X_2BH_2BX_2$, in which such multiple bonding would be slight. In the trihalogenoboranes BX_3, the average π-bond energy per $B \rightleftharpoons X$ bond has been estimated to be as follows:[19]

B–F, 68 kJ/mole; B–Cl, 42 kJ/mole; B–Br, 36 kJ/mole

If all this π-bond energy were lost on dimerization of dihalogenoboranes, the loss of π-bond energy associated with the process

$$2BHX_2 \rightarrow X_2BH_2BX_2$$

would be as follows:

for X=F, 272 kJ/mole; for X=Cl, 168 kJ/mole; for X=Br, 154 kJ/mole

All three exceed the enthalpy of dimerization of diborane (about 150 kJ/mole). Since there would also be a decrease in entropy on dimerization as the system became more ordered, the free energy change on dimerization would be positive ($\Delta G = \Delta H - T\Delta S$). The lack of association of these dihalogenoboranes is thus intelligible.

Alternatively, it has been pointed out that hydrogen bridges of the type that occurs in diborane are found only in hydridic species, species in which hydrogen is at the *negative* end of a bond attached to a more electropositive atom or group. The two halogen atoms on boron in a dihalogenoborane HBX_2, would by their combined inductive effects so drain electronic charge from boron, and in turn from hydrogen, as to make the latter positively charged relative to boron, and so no longer suitable for an electron deficient bridging role.

Halogen atoms do not normally make use of their lone pair electrons to form bridges B–X→B between boron atoms, and so the monohalogenodiboranes B_2H_5X have terminal halogens. In contrast, aminodiboranes $B_2H_5NR_2$, prepared by the action of an excess of diborane on a secondary amine at about 90° to ensure hydrogen elimination, have bridging nitrogen atoms, and so are known as μ-aminodiboranes:

The [11]B n.m.r. spectra of these μ-aminodiboranes consist of a triplet of doublets, reflecting [11]B–[1]H$_{terminal}$ and [11]B–[1]H$_\mu$ spin–spin coupling, when the spectra are recorded at low temperature. As the temperature is increased, and the terminal and bridging hydrogen atoms exchange places increasingly readily, the [11]B n.m.r. spectrum changes form to that appropriate for boron atoms coupling equally to five hydrogens, a sextet of relative intensities 1:5:10:10:5:1 (see Fig. 4–10).[20] The scrambling of the hydrogen

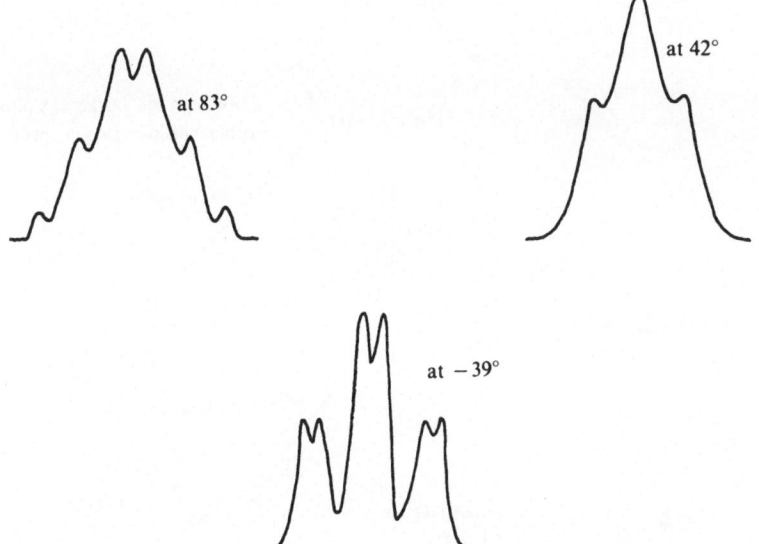

Fig. 4-10 The 19·3 MHz ^{11}B n.m.r. spectrum of μ-dimethylaminodiborane, $B_2H_5NMe_2$, at different temperatures in ethylene glycol dimethyl ether solution

atoms probably involves a mechanism common to many bridge species, alternate BHB bridge breaking and making:

(this BH_2H^* unit can rotate about the $N \rightarrow B$ bond)

$$H_2B \underset{N}{\overset{H^*}{\diamondsuit}} BH_2 \rightleftharpoons H_2B \underset{\underset{R_2}{N}}{\overset{H^*}{\diamondsuit}} BH_2 \rightleftharpoons H_2B \underset{\underset{R_2}{N}}{\overset{H}{\diamondsuit}} BHH^*$$

Some indication of the variety of products that can be prepared from diborane and ammonia with or without elimination of hydrogen is given in Fig. 4–11. Few of these reactions lead to essentially only one product— boron–nitrogen oligomers or polymers of ill-defined composition are a normal by-product, if not the major product, of reactions in which diborane and ammonia are heated together or allowed to react rapidly when heat is generated spontaneously. Interestingly, the product of symmetrical cleavage of the BH_2B bridge of diborane, the ammonia-borane adduct H_3N,BH_3, though not obtained by the direct action of ammonia on diborane, can be prepared by the reaction between an ammonium salt and a tetrahydroborate. Evidently ammonium tetrahydroborate loses hydrogen too readily to be isolated at room temperature.

$$NH_4Cl + LiBH_4 \xrightarrow{-LiCl} [NH_4BH_4] \xrightarrow{-H_2} H_3N,BH_3$$

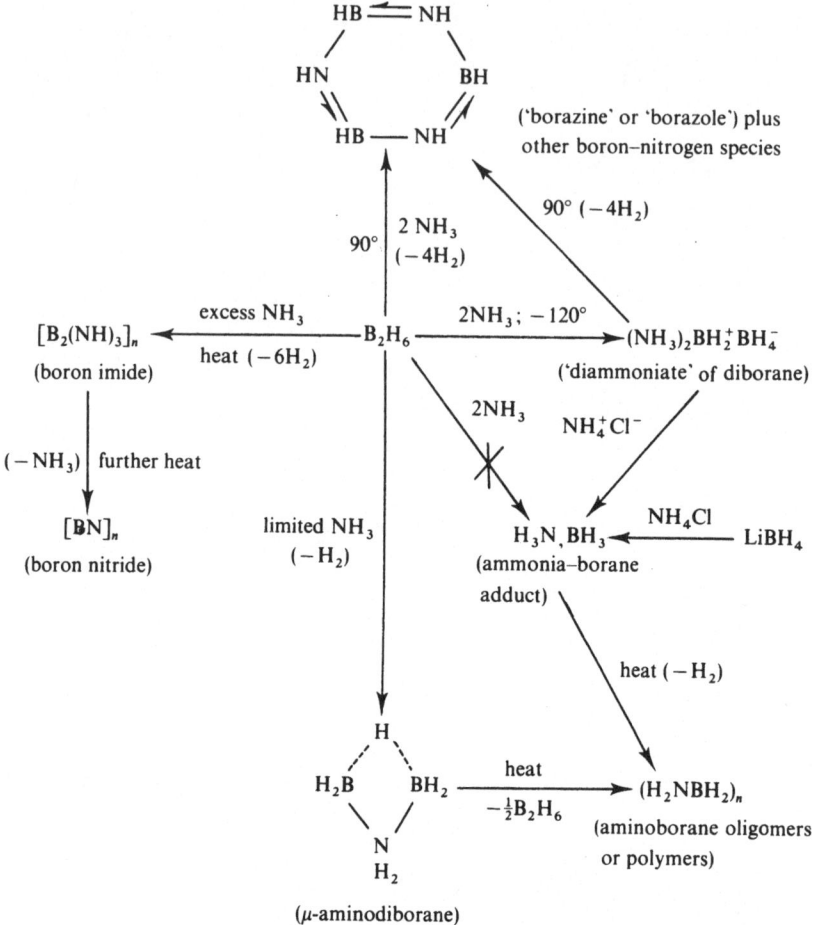

Fig. 4–11   The types of compound that can be prepared by reactions between diborane and ammonia.

A major product of reactions in which the diborane–ammonia ratio is $B_2H_6:2NH_3$, and in which the temperature is high enough (at least 90°C) for hydrogen evolution, is borazine, $(HNBH)_3$, the boron–nitrogen analogue of benzene which like benzene has a six-membered planar ring structure containing three coordinate boron and nitrogen atoms (see Fig. 4–11). Dative π-bonding from nitrogen to boron contributes to the stability of this molecule. Puckered six-membered or planar four-membered boron–nitrogen rings feature in the structures of aminoboranes $(H_2NBH_2)_n$ or derivatives in which two hydrogen atoms or other substituents remain attached to each boron and nitrogen, which are accordingly four co-ordinate. Eight-membered $(BM)_4$ rings are common in the products of reactions of phosphines or arsines R_2MH with diborane, for example

$$B_2H_6 + 2Me_2MH \rightarrow 2Me_2MH,BH_3 \xrightarrow[-2H_2]{heat} (Me_2MBH_2)_{3 \text{ or } 4}$$
$$(M = P,As)$$

Reactions of this type have been studied (as yet with little success) as possible routes to useful boron–nitrogen or boron–phosphorus polymers or polymer precursors, which could prove more thermally stable than analogous organic polymers.

Apart from their use for the synthesis of new compounds, hydrogen cleavage reactions of diborane and its derivatives have considerable analytical value. Controlled quantitative hydrolysis with subsequent volumetric measurement (in the vacuum line or in a gas burette or syringe) of the hydrogen evolved allows accurate determination of the B–H groups in a derivative of diborane, or of the concentration of a solution of diborane. Alcoholysis using for example 2-methoxyethanol $MeOCH_2$-CH_2OH added to a sample at $-78°$ or below is often convenient when direct hydrolysis proves too vigorous:

$$B_2H_6 + 6MeOC_2H_4OH \rightarrow 2B(OC_2H_4OMe)_3 + 6H_2$$

3   Addition of B–H to unsaturated groups—hydroboration[13]

When diborane is bubbled in a stream of nitrogen through a solution of an olefin in an ethereal solvent, organoboranes are formed:

$$6RCH:CH_2 + B_2H_6 \rightarrow 2(RCH_2CH_2)_3B$$

Any unchanged diborane in the exit gases may be safely removed by passing them through a bubbler containing acetone:

$$6Me_2C:O + B_2H_6 \rightarrow 2(Me_2CHO)_3B$$

Both these reactions are examples of hydroborations, reactions in which the B–H groups of diborane add across unsaturated groups. H. C. Brown and his coworkers[13] have shown the hydroboration of olefins to have considerable value in synthetic organic chemistry, because the organoboranes formed can by suitable treatment (see Table 4–6) be converted into a wide range of functionally substituted products. The hydroboration of other unsaturated groups (for example $R_2C:O$, $RC:N$, $RN:O$, etc.) containing hetero atoms also has its uses, in that hydrolysis of the product achieves overall hydrogenation (see Table 4–7), for example,

$$R^1R^2C:O \xrightarrow{B_2H_6} (R^1R^2CHO)_3B \xrightarrow{H_2O} R^1R^2CHOH$$

The orientation of hydroboration of olefins is *cis*-anti-Markownikov,

Table 4-6 Useful reactions with which to work up the organoboranes produced by the hydroboration of olefins

$$R^1-b^* + R^2CO_2H \rightarrow R^1H$$

$$R^1CH_2CH_2-b + R^2CH:CH_2 \rightarrow R^1CH:CH_2$$

$$R-b + H_2O_2/OH^- \rightarrow ROH$$

$$R-CH_2-b + H_2CrO_4 \rightarrow RCO_2H$$

$$R^1R^2CH-b + H_2CrO_4 \rightarrow R^1R^2CO$$

$$R-b + AgNO_3/OH^- \rightarrow R-R$$

$$R-b + Et_2NCl \rightarrow RCl$$

$$R-b + I_2/OH^- \rightarrow RI$$

$$R-b + BrCH_2COX/Bu^tOK/Bu^tOH \rightarrow RCH_2COX \qquad (X = R^1 \text{ or } OR^1)$$

$$R^1-b + CH_2:CHCOR^2 \rightarrow R^1CH_2CH_2COR^2$$

$$R-b + NH_2X \rightarrow RNH_2 \qquad (X = Cl \text{ or } OSO_3H)$$

$$R_3B + CO \xrightarrow[\text{(ii) OH}^-]{\text{(i) BH}_4^-/\text{ether solvent}} RCH_2OH$$

$$R_3B + CO \xrightarrow[\text{(ii) H}_2O_2/\text{NaOH}]{\text{(i) wet diglyme}/100^\circ} R_2CO$$

$$R_3B + CO \xrightarrow[\text{(ii) H}_2O_2/\text{NaOH}]{\text{(i) dry diglyme}/100^\circ} R_3COH$$

*b = 1/3 B

i.e. the boron becomes attached to the less substituted, relatively negatively charged, carbon of the olefin: a four-centre mechanism appears likely:

Ethylene itself and terminal olefins $R^1R^2C:CH_2$ rapidly react with all the B–H bonds of diborane even at room temperature, forming trialkylboranes $(R^1R^2CHCH_2)_3B$, which are monomeric; alkyl groups, unlike hydrogen atoms, are not known to form electron deficient bridges between boron atoms. Steric hindrance may cause the reaction to stop at an earlier stage for highly substituted olefins. For example, trimethylethylene forms the di-isoamylborane ('disiamylborane') dimer,

$$Me_2C:CHMe \xrightarrow{B_2H_6} [(Me_2CHCHMe)_2BH]_2$$

while tetramethylethylene rapidly adds to only one B–H group per boron, forming the dimeric hexylborane $(Me_2CHCMe_2BH_2)_2$ ('thexylborane').

Table 4-7 Reduction of organic functional groups by diborane

Reactant	Product	Reactant	Product
$R^1R^2C{:}CR^3R^4$	$R^1R^2CHCHR^3R^4$	$RN{:}NR$	RNH_2
$R^1C{:}CR^2$	cis-$R^1CH{:}CHR^2$	$RC{:}N$	RCH_2NH_2
$RC{:}OX^*$	RCH_2OH	$R^1R^2C{:}NR^3$	$R^1R^2CHNHR^3$
$R^1R^2C{:}O$	R^1R^2CHOH	$R^1R^2C{:}NX^*$	$R^1R^2CHNH_2$
$R\overline{CHCH_2O}$	RCH_2CH_2OH	$RN{:}O$	RNH_2
$\overline{R.CO.O}$	$HORCH_2OH$	$R_2S{:}O$	R_2S

A hydrolytic procedure for working up the product is needed in each case.
$^*X = H, OH, OR, Cl, CO_2R$

Although these alkyldiboranes are so crowded as to react further with their parent olefins only with difficulty, their B–H groups are capable of adding to less hindered olefins, which they do in a highly stereospecific manner, for example,

$$PhCH{:}CH_2 + (R_2BH)_2 \rightarrow 98\% \ PhCH_2CH_2BR_2 (R = isoamyl)$$

cf. $PhCH{:}CH_2 + B_2H_6 \quad \rightarrow 80\% \ PhCH_2CH_2B\diagup_\diagdown + 20\% \ PhCH(Me)B\diagup_\diagdown$

Disiamylborane dimer and thexylborane dimer can be used to selectively hydroborate one olefin in the presence of another of a different structural type. Treatment of commercial 2-pentene (which consists of 18% *cis*- and 82% *trans*-isomer) selectively hydroborates the *cis*-isomer, so leaving essentially pure *trans*-isomer.

Another reagent which has proved useful for selective hydroboration reactions is 9-borabicyclo[3.3.1]-nonane ('9–BBN'), prepared by the hydroboration of cyclo-octa-1,5-diene:

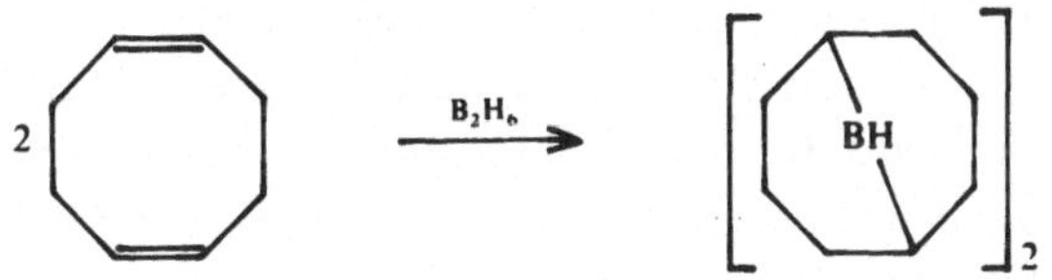

Optically stereospecific hydroboration can be achieved by use of di-isopinocampheylborane dimer (from α-pinene and diborane):

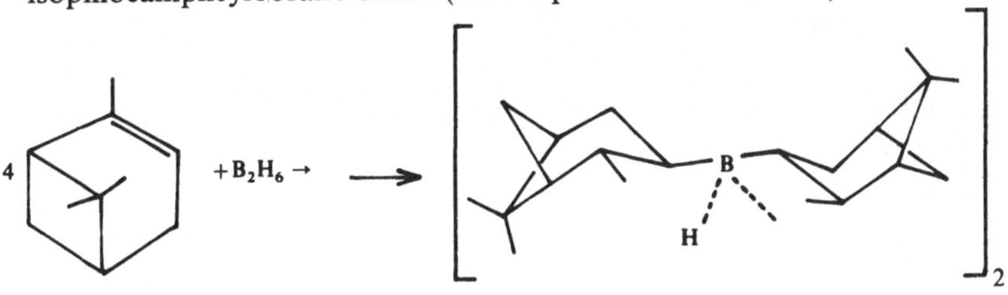

Studies with this reagent have shown that its *cis-*, anti-Markownikov addition to olefins probably involves the *dimer* (the reaction is first order in both olefin and the di-isopinocampheylborane *dimer*). The suggested mechanism involves approach of the olefin along the two-fold axis through the bridging hydrogens with an orientation appropriate for a bonding interaction between the olefin π-bonding orbital and the vacant $1b_{2g}$ $\pi*$ orbital of the diborane, as in Fig. 4–12.[21] The olefin subsequently

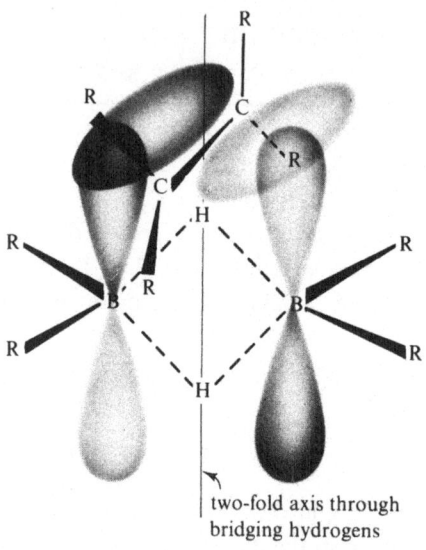

two-fold axis through
bridging hydrogens

Fig. 4-12   Possible mode of approach of an olefin to a tetra-alkyldiborane along the two-fold axis through the bridging hydrogen atoms, with a suitable orientation to act as a π-donor to the vacant $1b_{2g}$ diborane $\pi*$ orbital

twists down on to one of the bridging B–H units in the boron–carbon σ-bond-making stage of the reaction; in this stage, its orientation is governed both by its own substituents (the least substituted carbon dips down towards the boron) and also by the shape and chirality of the di-isopinocampheyl substituents on the diborane. This mechanism allows the chiralities of products derived from *cis-*1,2-dialkyl olefins $R^1CH:CHR^2$ to be correctly deduced, and is consistent with the detailed stereochemistry of many other systems, though alternative competing mechanisms remain possible in some. A unique mechanism is hardly to be expected in systems containing such complicated mixtures of potentially reactive species (B_2H_6, R_2O,B_2H_6, R_2O,BH_3, possibly $(R_2O)_2BH_2^+BH_4^-$, and alkyl derivatives of these). The *cis*-addition of B–H to acetylenes may occur similarly.

In the hydroboration of functional groups such as ketones and nitriles, σ-bonded complexes L,B_2H_6 or L,BH_3 ($L = R_2C:O$ or $RC\vdots N$)

are probably involved: nitrile adducts $RC\!:\!N,BH_3$ have been isolated and characterized, though they rearrange spontaneously below room temperature. A likely mechanism involves six-membered ring intermediates:

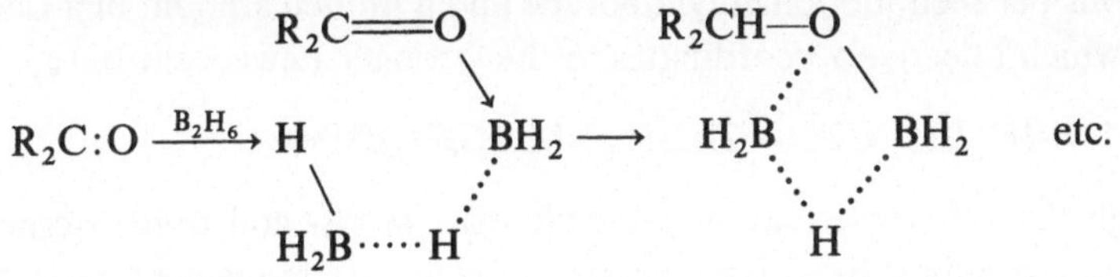

4 Exchange reactions

Reactions between diborane and other compounds BX_3 can be used to prepare substituted diboranes B_2H_5X, $B_2H_4X_2$, etc., without the wastage of links to boron that occurs in reactions between diborane and protic acids HX. That reactions with trialkylboranes R_3B gave derivatives B_2H_5R, $B_2H_4R_2$, $B_2H_3R_3$, and $B_2H_2R_4$, but not B_2HR_5 or B_2R_6, provided early evidence that two of the hydrogen atoms of diborane differed from the other four; only the terminal hydrogens can be replaced by alkyl groups, though the latter probably adopt a bridging role during transfer:

for example,

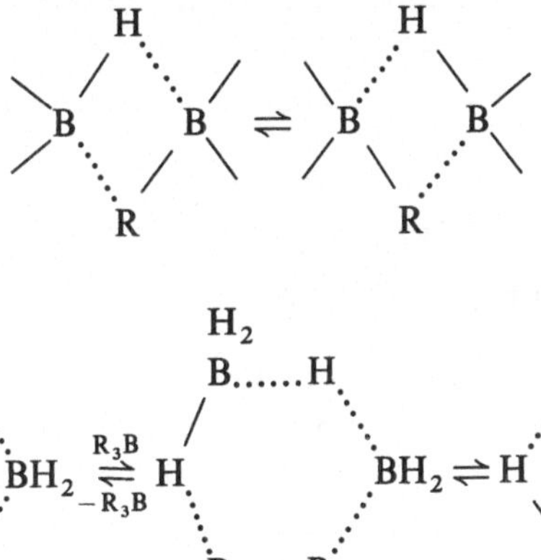

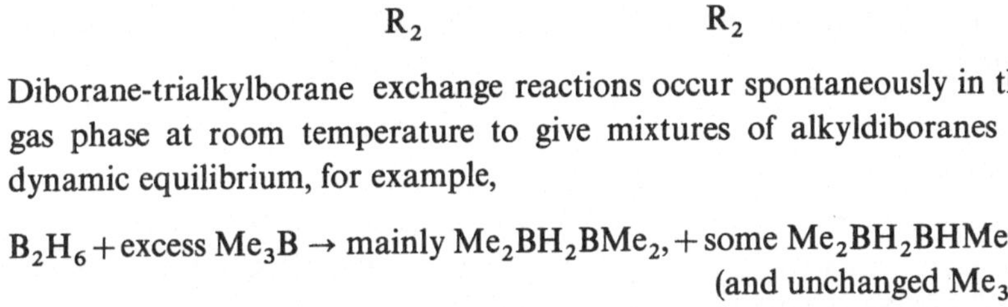

or

Diborane-trialkylborane exchange reactions occur spontaneously in the gas phase at room temperature to give mixtures of alkyldiboranes in dynamic equilibrium, for example,

$$B_2H_6 + \text{excess } Me_3B \rightarrow \text{mainly } Me_2BH_2BMe_2, + \text{some } Me_2BH_2BHMe$$
$$(\text{and unchanged } Me_3B)$$

Equimolar proportions of diborane and trimethylborane give predominantly the 1,1-dimethyl derivative $Me_2BH_2BH_2$. Its ^{11}B n.m.r. spectrum contains a singlet due to the boron of Me_2B, a triplet due to the other boron. Moreover, on hydrolysis it gives dimethylborinic acid Me_2BOH and boric acid $B(OH)_3$. The 1,2 isomer $MeHBH_2BHMe$ (which gives methylboronic acid, $MeB(OH)_2$, on hydrolysis) results from reactions between monomethyldiborane and a limited amount of a Lewis base, which selectively coordinates to the stronger Lewis acid BH_3:

$$2MeHBH_2BH_2 + 2L \rightarrow 2L,BH_3 + MeHBH_2BHMe$$

Though the 1,2-isomer can in principle exist in cis- and trans- isomeric forms, these cannot be isolated because of facile rearrangement, probably via an intermediate with a single hydrogen bridge in which the two sections of the molecule can rotate with respect to each other:

$$
\begin{array}{ccc}
\underset{H}{\overset{Me}{\diagdown}}BH_2B\underset{H}{\overset{Me}{\diagup}}
& \rightleftharpoons
\left[\underset{H}{\overset{Me}{\diagdown}}H\!\!-\!\!BHB\underset{H}{\overset{Me}{\diagup}} \right]
& \rightleftharpoons
\underset{H}{\overset{Me}{\diagdown}}BH_2B\underset{Me}{\overset{H}{\diagup}}
\end{array}
$$

Exchange reactions between diborane and boron trihalides lead to the same products as can be prepared also by diborane–hydrogen halide reactions. Exchanges with the halides or alkoxides of other elements lead to the hydride or tetrahydroborate of the other element, though a better route to these involves the use of alkali metal tetrahydroborates, for example,

$$(Me_3Al)_2 + 4B_2H_6 \rightarrow 2Al(BH_4)_3 + 2Me_3B$$

cf. $AlCl_3 + 3NaBH_4 \rightarrow Al(BH_4)_3 + 3NaCl$

The self-exchange reactions of diborane (for example, between $B_2{}^1H_6$ and $B_2{}^2H_6$) and the exchange with molecular deuterium, which generate species $B_2{}^1H_n{}^2H_{6-n}$ ($n = 0 \rightarrow 5$) are believed to involve the same intermediates as feature in the thermal decomposition of diborane (see Section 4–3, above).[22,23] The self-exchange reaction for example is of order $\frac{3}{2}$ in diborane, indicating that triborane(9), B_3H_9, is a probable intermediate.

5   Oxidative cleavage of B–H bonds

Reactions of this type $(B–H + X_2 \rightarrow B–X + HX)$ are the source of the oxygen-sensitivity of diborane and other boron hydrides, but are rarely of value for the synthesis of specific boron compounds, being wasteful of boron–hydrogen links.

Much attention has been paid to the combustion of diborane to boric oxide and steam, particularly to such features as explosion limits, flame velocities, etc., that relate to the use of diborane as a fuel. The reaction apparently is exceedingly complicated, proceeds by a radical chain mechanism, and produces too many ill-defined products of incomplete combustion to be worth exploiting in rocket engines:

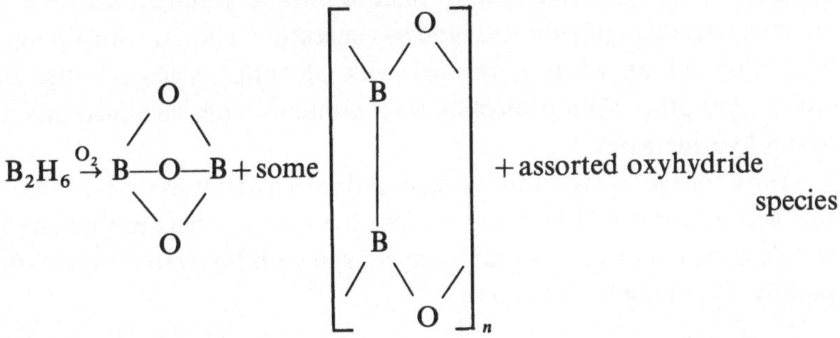

6   Reduction of diborane

The addition of electrons to diborane using alkali metal or alkaline earth amalgam leads to the formation of an equimolar mixture of the hydroborate anions BH_4^- and $B_3H_8^-$.

$$2NaHg + 2B_2H_6 \rightarrow NaBH_4 + NaB_3H_8 + 2Hg$$

This reaction may involve such ionic species as $B_2H_6^-$ or $B_2H_6^{2-}$ as intermediates, though neither has been isolated or even detected. The hexachlorodiborate $(2-)$ anion, $[Cl_3B-BCl_3]^{2-}$ (cf. hexachloroethane), is known as the product of reactions between diboron tetrachloride and chloride donors:

$$Cl_2B-BCl_2 + 2Cl^- \rightarrow Cl_3B-BCl_3^{2-}$$

As a route to the octahydrotriborate anion, $B_3H_8^-$, the reduction of diborane with sodium amalgam is less convenient than the reaction with tetrahydroborate in diglyme, $(MeOCH_2CH_2)_2O$, at 100°

$$NaBH_4 + B_2H_6 \rightarrow NaB_3H_8 + H_2$$

The product is the material from which the icosahedral *closo-* anion $B_{12}H_{12}^{2-}$ can be prepared:[24-26]

$$5B_3H_8^- \xrightarrow{\text{heat}} B_{12}H_{12}^{2-} + 3BH_4^- + 8H_2$$

7   Decomposition to hydrogen and higher hydrides

The thermal decomposition of diborane as a route to higher hydrides was described in section 4–3. Its decomposition in the presence of acetylenes leads to mixtures of alkylboranes and carboranes (see Chapter 6).

4-5 Some reactions of higher boranes

Tetraborane(10), B_4H_{10}[5,9-12]

Tetraborane(10) was the source from which Stock prepared his other boron hydrides by thermal decomposition. He described it as having an even more disagreeable odour than diborane (though smelling like chocolate when highly diluted) and as generating a liquid with a pungent, disgusting odour when it reacted with alcohol.[2] Vacuum lines have advantages other than protecting their contents from chemical attack in boron hydride work.

Tetraborane (10) is much less susceptible to hydrolysis than diborane—this was why it could be isolated from the reaction between magnesium boride and aqueous acid—but when shaken with hot water it reacts quite rapidly according to the equation:

$$B_4H_{10} + 12H_2O \rightarrow 4H_3BO_3 + 11H_2$$

It reduces aqueous permanganate solutions to manganese dioxide, and is oxidized explosively by concentrated nitric acid. Like diborane, it is liable to inflame on exposure to air. Studies on its interaction with a limited amount of oxygen have shown that a chain mechanism probably operates, involving various boron oxyhydrides as intermediates. The precise composition of these has been difficult to establish, but among the products that have been characterized are substances containing BH units linked by oxygen bridges:

$$B_4H_{10} + \text{limited } O_2 \rightarrow \text{ some }
\begin{array}{c}
\text{HB} \quad\quad\quad \text{HB---O} \\
/ \quad \text{O} \quad\quad / \quad\quad \text{\textbackslash} \\
| \quad + \text{O} \quad\quad\quad \text{BH} \\
\text{\textbackslash} \quad \text{O} \quad\quad \text{\textbackslash} \quad / \\
\text{HB---} \quad\quad \text{HB---O}
\end{array}$$

Short-lived species such as HBO, H_2BOH, and H_3BO_2 may be involved as intermediates.

Like diborane, tetraborane(10) has a structure in which there are two BH_2 units, each of which is linked to the rest of the molecule by two BHB bridges. This structural similarity is reflected in the types of reaction it undergoes. For example, it is susceptible to attack by nucleophiles; Lewis bases cleave the BHB bridging units in a manner reminiscent of their effect on diborane (see Fig. 4–13). Note that, as with diborane, when $L = H^-$ the same products are obtained whether attack occurs successively at the same boron atom or at different borons. These reactions are believed to proceed via initial attack on boron-2 or -4 (the borons of the BH_2 groups), which are slightly more positively charged than the other

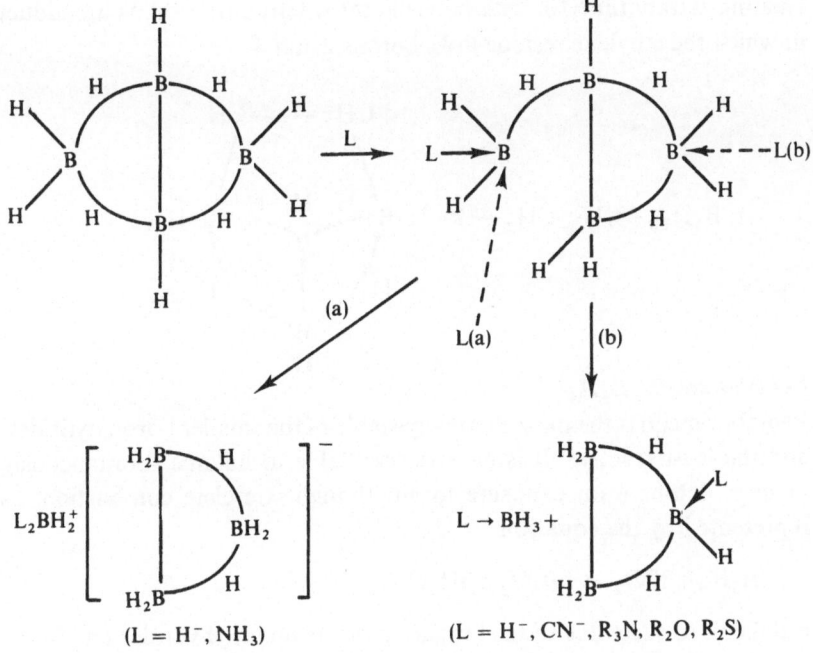

Fig. 4-13   Reactions between tetraborane (10) and Lewis bases

borons, the important difference being that they have a terminal $B(\delta+)$—$H(\delta-)$ group where borons -1 and -3 have a B—B link.

The bridging and terminal hydrogen atoms of the adducts L,B_3H_7 exchange positions readily, apparently because alternative structures are possible:

With concentrated alkali, tetraborane(10) forms a mixture of borate and hydroborates:

$$2B_4H_{10} + 4KOH \rightarrow 2KB_3H_8 + KBH_4 + KB(OH)_4'$$

Tetraborane(10) decomposes slowly at room temperature, rapidly at 60°C, when pentaborane(11) is a major product. Loss of hydrogen to form the highly reactive intermediate tetraborane(8) is among the reactions probably involved:

$$B_4H_{10} \rightleftharpoons B_4H_8 + H_2$$

Heating tetraborane(10) with ethylene traps tetraborane(8) as an adduct in which the ethylene residue links borons 2 and 4:

$$B_4H_{10} + CH_2{:}CH_2 \longrightarrow$$

Pentaborane(9), B_5H_9 [5,9-12]

Pentaborane(9) is the most thermally stable of the smaller boron hydrides, and the least reactive. It is nevertheless liable to inflame spontaneously or even detonate on exposure to air, though complete combustion, as represented by the equation

$$2B_5H_9 + 12O_2 \rightarrow 5B_2O_3 + 9H_2O$$

is difficult to effect due to the formation of various oxyhydrides and lower oxide species. It is only very slowly hydrolyzed by water, with which it is

Table 4-8　Some reactions of pentaborane(9)

---

1 Electrophilic substitution of terminal hydrogen

$$B_5H_9 + X_2 \rightarrow 1\text{-}XB_5H_8 \qquad (X = Br, I)$$

$$B_5H_9 + Cl_2/AlCl_3 \xrightarrow{0°} 1\text{-}ClB_5H_8$$

$$B_5H_9 + RX/AlCl_3 \rightarrow 1\text{-}RB_5H_8 \qquad (RX = MeI, EtBr, PrCl, BuCl)$$

$$B_5H_9 + {}^2HCl/AlCl_3 \rightarrow 1\text{-}{}^2HB_5H_8$$

2 Rearrangement of substituents

$$1,2\text{-}Cl_2B_5H_7 \xrightarrow{\text{hexamethylenetetrammine}} 2,3\text{- and } 2,4\text{-}Cl_2B_5H_7$$

$$1\text{-}{}^2HB_5H_8 \xrightarrow[\text{intramolecular exchange } 0°]{2,6\text{-dimethylpyridine}; \ 20°} 2\text{-}{}^2HB_5H_8 + \mu\text{-}{}^2HB_5H_8$$

3 Reactions with olefins and acetylenes

$$B_5H_9 + \text{olefins} \longrightarrow 2\text{-}RB_5H_8 \qquad \text{(base substitution)}$$

$$B_5H_9 + \text{olefins} \xrightarrow{AlCl_3} 1\text{-}RB_5H_8 \qquad \text{(apical substitution)}$$

$$B_5H_9 + \text{acetylenes} \xrightarrow{\text{base}} \text{derivatives of } B_4C_2H_8$$

$$B_5H_9 + \text{acetylenes} \xrightarrow[\text{or electric dicharge}]{\text{strong heat, u.v.,}} B_3C_2H_5, B_4C_2H_6, B_5C_2H_7, \text{etc.}$$
$$\text{(carboranes)}$$

4 Exchange reactions with diborane

$${}^{11}B_5H_9 \xrightarrow{{}^{10}B_2H_6; \ Al_2O_3} {}^{10}B^{11}B_4H_9$$

$$B_5{}^1H_9 \xrightarrow{B_2{}^2H_6} B_5{}^1H_4{}^2H_5$$

---

immiscible, though when a common solvent for the reactants is added (for example, dioxane), rapid reaction occurs. Alcohols overnight generate 12 moles of hydrogen per mole of pentaborane(9):

$$B_5H_9 + 15ROH \rightarrow 12H_2 + 5B(OR)_3$$

With less alcohol, alkoxyboranes $ROBH_2$ or $(RO)_2BH$ may result.

Some typical reactions of pentaborane(9) are listed in Table 4–8. The relatively negative charge on the apical boron, boron-1, makes this boron susceptible to electrophilic attack. Chlorination, alkylation, or deuteration at this boron can be effected under Friedel–Crafts conditions. Subsequent rearrangement of substituents from the apical to a basal boron occurs in the presence of bases, which apparently function by abstracting, as a proton, one of the bridging hydrogens. This leaves an anion $B_5H_7R^-$ which can easily rearrange by migration of a hydrogen from its basal BH_2 group to what was originally the apical boron (see Fig. 4–14).

Pentaborane(9) reacts with olefins to form alkyl derivatives; attack occurs on basal borons in the absence of catalyst, on the apical boron in the presence of aluminium chloride. With acetylenes, pentaborane(9) reacts

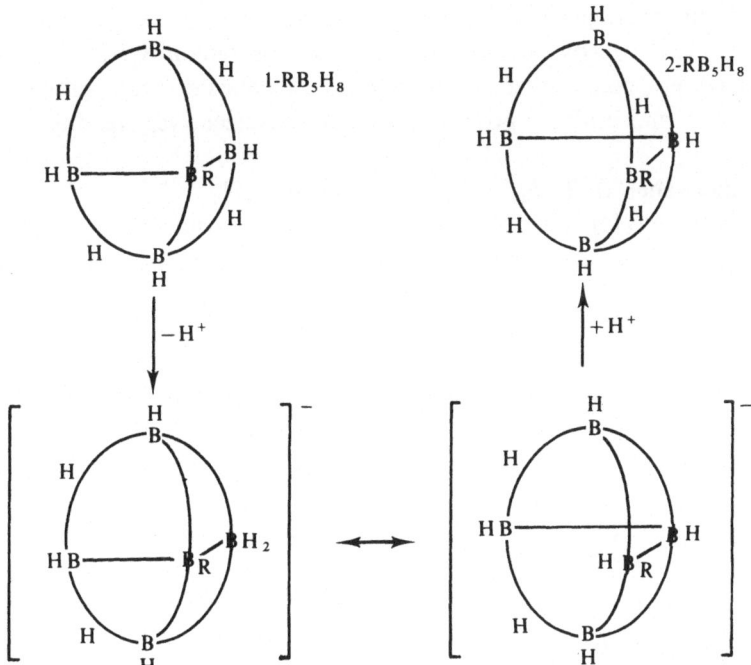

Fig. 4–14 A mechanism for the base-catalyzed rearrangement of 1-RB_5H_8 to 2-RB_5H_8 via the anion $[RB_5H_7]^-$

under high energy conditions to form products in which carbon as well as boron atoms are present in the electron deficient molecular skeleton (Table 4–8). These products, known as carboranes, are discussed in Chapter 6.

The exchange of labelled boron or hydrogen between pentaborane(9) and diborane has been studied to get more information on the inter-conversion of these key boron hydrides. No exchange of boron atoms between $^{11}B_5H_9$ and $^{10}B_2H_6$ occurs below the decomposition tempera-ture of the former unless a catalyst such as alumina is present, when apparently one boron atom of the pentaborane is replaced by the ^{10}B isomer, within 30 minutes at 90°C. Exchange of hydrogen atoms between these substances occurs more readily. Pentaborane(9) can be deuterated by means of $B_2{}^2H_6$ in a reaction which is first order in B_5H_9 and half-order in $B_2{}^2H_6$, a result taken to show that B^2H_3 may be the effective reagent. Interestingly, at 80°C during 24 hours in the absence of catalyst, only five of the hydrogen atoms of pentaborane(9) are deuterated. These are the five terminal hydrogen atoms, as shown by the shift of $\nu(B–H)$ in the infrared spectrum from 2600 cm^{-1} to 1960 cm^{-1}, and the loss of $(^{11}B–^1H_{terminal})$ spin–spin coupling fine structure in the ^{11}B n.m.r. spec-trum. This shows not only that it is the terminal hydrogen atoms of B_5H_9 that exchange with those of diborane, but also that terminal ↔ bridging hydrogen rearrangements do not occur within neutral B_5H_9 (contrast the ready exchange that occurs in the anion $B_5H_8^-$, Fig. 4–14, R=H). For many other boron hydrides, exchange of hydrogen between terminal and bridging positions is believed to occur very readily.

Pentaborane(11), B_5H_{11} and hexaborane(10), B_6H_{10}[5,9-12]

Pentaborane(11) is a highly reactive, thermally unstable hydride. Many of its reactions reflect the ready reversible loss of monoborane(3):

$$B_5H_{11} \rightleftharpoons [BH_3] + [B_4H_8]$$

For example, with carbon monoxide it forms the adducts $H_3B{,}CO$ and $B_4H_8{,}CO$ and with ethylene it gives the same product, $C_2H_4B_4H_8$ as is formed by tetraborane(10), together with ethylpentaborane(11):

$$B_5H_{11} \rightleftharpoons [BH_3] + [B_4H_8] \xrightarrow{C_2H_4} C_2H_4B_4H_8$$

$$\Big\downarrow C_2H_4 \qquad \overset{[EtBH_2]}{\underline{\qquad\qquad}} \to EtB_5H_{10}$$

$$[EtBH_2]$$

Lewis bases such as amines give complex reactions with extensive breakdown of the molecule. Ethers catalyse the formation of hexaborane

(10), diborane and tetraborane(10). This last is formed quantitatively in the partial hydrolysis of pentaborane(11):

$$B_5H_{11} + 3H_2O \rightarrow B_4H_{10} + H_3BO_3 + 2H_2$$

Pentaborane(11) readily exchanges both boron and hydrogen with diborane in a reaction first-order in B_2H_6, half-order in B_5H_{11}, evidently involving the action of $[BH_3]$ and $[B_4H_8]$ on diborane as rate-determining steps.

Hexaborane(10) is less readily prepared by the thermal decomposition of diborane than are the other hydrides, and its chemistry has accordingly been little studied. It is incompletely hydrolysed by water at 90° or by 30% alkali at lower temperature, but is completely hydrolysed by acid:

$$B_6H_{10} + 18H_2O \xrightarrow{\text{acid}} 6B(OH)_3 + 14H_2$$

Decaborane(14)[5,9–12,27]

Decaborane(14) is the only solid common borane, and the least volatile. It resists oxidation sufficiently to be handled in air, and can be recovered unchanged from aqueous solution. It may, however, react explosively with halogenated solvents, which should therefore be avoided. After diborane, it is the borane whose chemistry has been most extensively studied. It is capable of a wide range of reactions some of which lead to derivatives $B_{10}H_{13}X$, $B_{10}H_{12}X_2$, etc., in which the molecular skeleton is retained and atoms or groups X replace terminal hydrogen atoms. Others lead to related hydride anions, for example, $B_{10}H_{13}^-$, $B_{10}H_{14}^{2-}$, $B_{10}H_{15}^-$, and $B_9H_{14}^-$ or derivatives thereof (see Table 4–9). Topological representations of the bonding in these species are shown in Fig. 4–15. Though the arrangement of the boron atoms in their skeletons differs little from one to another, there are significant variations in the numbers and positions of their terminal and bridging hydrogen atoms. The inter-relationships between the various types of compound that can be prepared from decaborane(14) are summarized in Fig. 4–16. Some representative reactions are listed in Table 4–10.

Table 4–9   **Key species in the chemistry of decaborane (14)**

Parent species	Isoelectronic (and isostructural) species
$B_{10}H_{14}$	$B_{10}H_{13}X$, $B_{10}H_{12}X_2$, $B_{10}H_{11}X_3$, etc.
$B_{10}H_{13}^-$	$B_{10}H_{12}X^-$, $B_{10}H_{12}L$
$B_{10}H_{14}^{2-}$	$B_{10}H_{13}L^-$, $B_{10}H_{12}L_2$
$B_{10}H_{15}^-$	$B_{10}H_{14}$,L
$B_9H_{14}^-$	B_9H_{13},L

X is an atom or group which, like a hydrogen atom, may be regarded as a one-electron donor

L is an electron-pair donor, and so equivalent to a hydride anion H^-

Table 4–10   Some reactions of decaborane (14)

Electrophilic substitution (at positions 1, 2, 3, or 4)

$B_{10}H_{14} \xrightarrow{I_2;\ Br_2;\ ICl/AlCl_3} B_{10}H_{13}X + B_{10}H_{12}X_2$

$B_{10}H_{14} \xrightarrow{RX/AlX_3} B_{10}H_{13}R,\ B_{10}H_{12}R_2,\ B_{10}H_{11}R_3,\ \text{and}\ B_{10}H_{10}R_4$

$B_{10}H_{14} \xrightarrow[CS_2]{^2HCl/AlCl_3} 1,2,3,4\text{-}B_{10}{}^1H_{10}{}^2H_4$

Nucleophilic substitution (6, 9 positions preferred; then 5, 7, 8, 10)

$B_{10}H_{14} \xrightarrow{RLi} B_{10}H_{13}R + B_{10}H_{12}R_2$

Reaction of $B_{10}H_{14}$ as a protic acid; formation of the anion $B_{10}H_{13}^-$ and derivatives.

$B_{10}H_{14} \underset{H_2O}{\rightleftharpoons} H_3O^+ B_{10}H_{13}^-$

$B_{10}{}^1H_{14} \xrightarrow[^2H_2O]{^2H^+} B_{10}{}^1H_{10}{}^2H_4 \xrightarrow[^2H_2O]{^2H^+} B_{10}{}^1H_4{}^2H_{10}$

$\qquad\qquad\qquad\qquad \vdots \qquad\qquad\qquad \vdots$

$\qquad\qquad\qquad\quad (\text{all } \mu) \qquad\quad (\mu, 5, 6, 7, 8, 9, \text{and } 10)$

$B_{10}H_{14} \xrightarrow[Et_2O]{NaH\ or\ LiBH_4} M^+ B_{10}H_{13}^- n Et_2O \xrightarrow{excess\ NaH} Na_2^+ B_{10}H_{12}^{2-}$

$B_{10}H_{14} \xrightarrow{L} LH^+ B_{10}H_{13}^- \qquad (L = NH_3,\ \text{amines},\ Ph_3PCH_2)$

$B_{10}H_{14} \xrightarrow{MeMgI} CH_4 + B_{10}H_{13}MgI,\ \text{the 'decaboranyl Grignard reagent'}$

$\qquad\qquad\qquad\qquad\qquad \searrow RX$

$\qquad\qquad\qquad\qquad 6\text{-}RB_{10}H_{13}$

Degradation to the anion $B_9H_{14}^-$

$B_{10}H_{14} \xrightarrow{aqueous\ OH^-} B_{10}H_{13}^- \rightarrow B_{10}H_{13}OH^{2-} \xrightarrow{H_3O^+} B_9H_{14}^-$

Reduction to the anions $B_{10}H_{14}^{2-}$ and $B_{10}H_{15}^-$

$B_{10}H_{14} \xrightarrow{Na/liquid\ NH_3} B_{10}H_{14}^{2-} \xrightarrow[H_2O]{H_3O^+} B_{10}H_{15}^-$

$B_{10}H_{14} \xrightarrow[aq.\ solution]{KBH_4} B_{10}H_{13}^- \rightarrow B_{10}H_{14}^{2-}$

Reactions attaching Lewis acids to borons -6 or -9.

$B_{10}H_{14} \xrightarrow[slow]{L} 6\text{-}B_{10}H_{14},L \xrightarrow[fast]{-H_2} B_{10}H_{12},L \xrightarrow[fast]{L} 6,9\text{-}B_{10}H_{12}L_2$

$\qquad\qquad\qquad (L = R_2S,\ RCN,\ R_2NCN,\ R_2NCOR,\ R_3N,\ R_3P,\ \text{pyridine})$

$B_{10}H_{14} + L \xrightarrow{R_2O} H_2 + B_{10}H_{12},L \qquad (L = CN^-,\ NCO^-,\ NCS^-,\ (NC)_3C^-)$

$B_{10}H_{14}^{2-} + 2LH^+ \rightarrow B_{10}H_{12}L_2 + H_2 \qquad (L = \text{pyridine})$

$B_{10}H_{14} \xrightarrow[-H_2]{NaH/Et_2O} B_{10}H_{13}^- \xrightarrow{L/Et_2O} B_{10}H_{13}L^-$

$\qquad\qquad\qquad (L = Et_2NH,\ EtNH_2,\ Et_3N,\ \text{pyridine, piperidine},\ Ph_3P,\ Me_2S)$

$B_{10}H_{14} + 2NaCN \xrightarrow{H_2O} HCN + Na_2^+ B_{10}H_{13}CN^{2-}$

$B_{10}H_{14} + Me_2S + CN^- \rightarrow H_2 + B_{10}H_{12}(SMe_2)CN^-$

$B_{10}H_{13}L^- + 2H_2O + H_3O^+ \rightarrow 2H_2 + B(OH)_3 + B_9H_{13}L$

$B_{10}H_{12}(SR_2)_2 + 3MeOH \rightarrow B_9H_{13},SR_2 + H_2 + B(OMe)_3 + R_2S$

$\qquad\qquad\qquad\qquad\quad \downarrow L$

$\qquad\qquad\qquad\qquad B_9H_{13},L + R_2S \qquad (L = RCN,\ R_2NH,\ Ph_3P,\ \text{pyridine})$

Formation of closed cage species

$$B_{10}H_{14} \xrightarrow[\text{or } R_3N \text{ strong heat}]{\text{(i) 2L, (ii) } R_3N} (R_3NH^+)_2 B_{10}H_{10}^{2-}$$

$$B_{10}H_{14} \xrightarrow{2L;\, RC\vdots CR} R_2C_2B_{10}H_{10}, \quad \text{1,2-disubstituted-1,2-dicarba-} \textit{closo-} \\ \text{dodecaboranes (carboranes)}$$

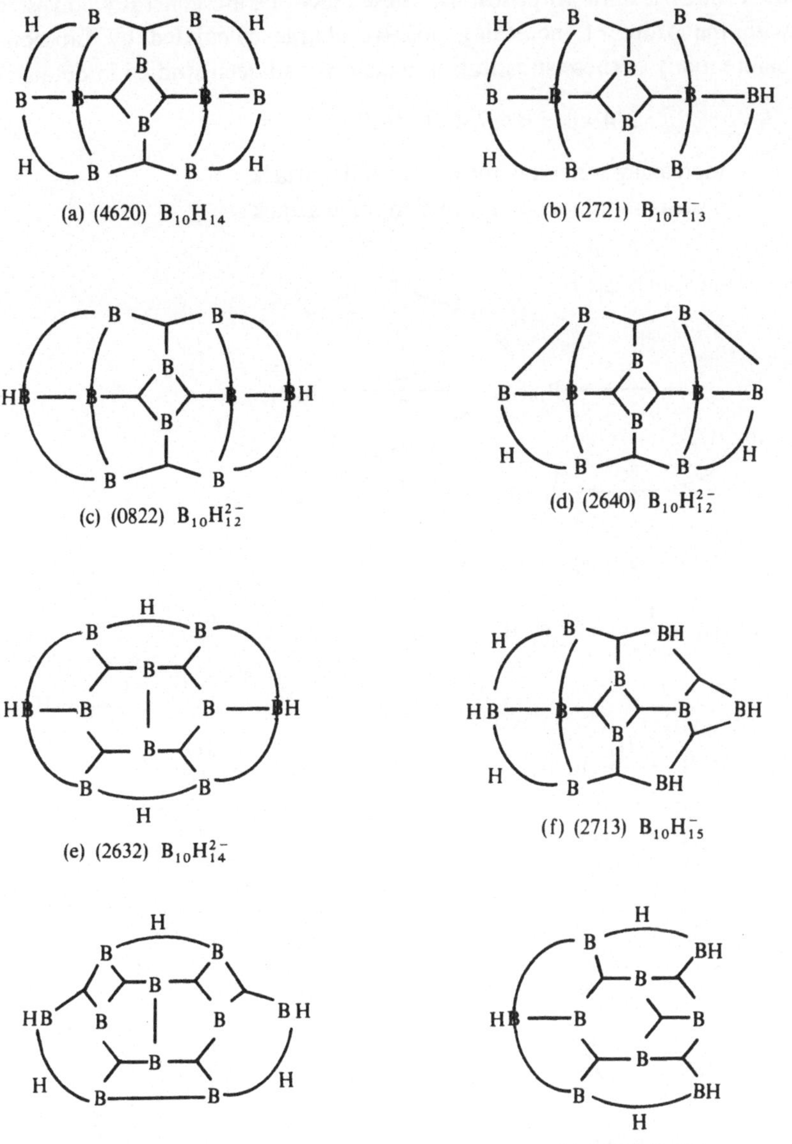

(a) (4620) $B_{10}H_{14}$

(b) (2721) $B_{10}H_{13}^-$

(c) (0822) $B_{10}H_{12}^{2-}$

(d) (2640) $B_{10}H_{12}^{2-}$

(e) (2632) $B_{10}H_{14}^{2-}$

(f) (2713) $B_{10}H_{15}^-$

(g) (3622) $B_{10}H_{15}^-$

(h) (2613) $B_9H_{14}^-$

(in all these structures, B represents the BH group)

Fig. 4-15 Topological representations and **styx** numbers of some key hydride species in the chemistry of decaborane(14)

Electrophilic and nucleophilic substitution of decaborane(14) occurs at the positions expected from the calculated charge distribution in the molecule (see Fig. 4–17). Thus electrophilic halogenation or Friedel–Crafts alkylation or deuteration effects substitution preferentially at borons 1 to 4, and carbanionic reagents preferentially substitute the 6 or 9 and then the 5, 7, 8, or 10 positions. These reactions are generally consistent with the order of increasing positive charge calculated by Lipscomb before many of these substitution reactions had been studied in detail, i.e.

$$2,4 < 1,3 < 5,7,8,10 < 6,9$$

←increasing tendency for electrophilic attack:
increasing tendency for nucleophilic attack :→

(a) Parent hydride species:

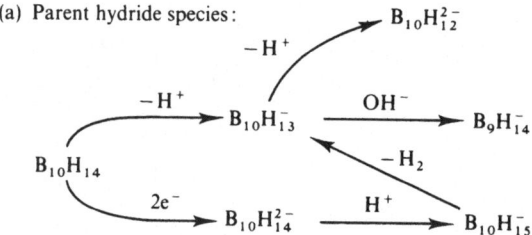

(b) Species incorporating Lewis bases L:

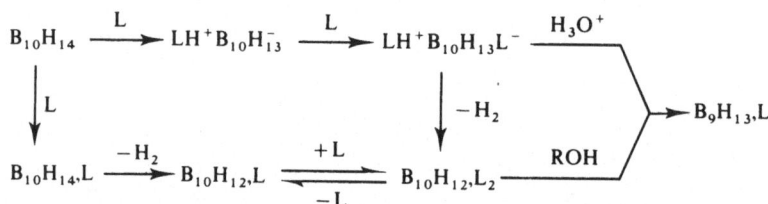

Fig. 4-16 Inter-relationships between some of the species preparable from decaborane(14)

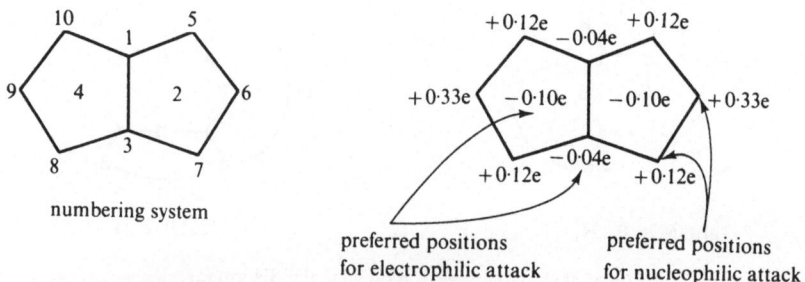

numbering system

preferred positions
for electrophilic attack

preferred positions
for nucleophilic attack

Fig. 4-17 Numbering system and calculated (exaggerated) charge distribution in decaborane(14)

In water and other polar solvents, decaborane(14) can be titrated as a monobasic acid, as it readily loses one of its bridging hydrogen atoms as a proton, forming the anion $B_{10}H_{13}^-$ which imparts a yellow colour to the solutions. This anion has a structure (Fig. 4–15b) with only two bridging hydrogen atoms—the other bridging hydrogen moves to a terminal position on boron-6, which thereby uniquely acquires altogether two terminally attached hydrogen atoms.

This ionization of a bridging hydrogen is reversible, and so allows deuteration, using deuterium oxide or deuterium chloride in dioxan, of the four equivalent bridging hydrogens of decaborane(14). Moreover, the terminal $\leftrightarrow$ bridging rearrangement of the remaining hydrogens of the anion $B_{10}H_{13}^-$ allows the hydrogens terminally attached to borons 6 and 9 to be deuterated also, and indeed ultimately all six terminal hydrogens adjacent to the bridging hydrogens can be deuterated thus, i.e. positions 5, 6, 7, 8, 9, and 10. The stepwise nature of these deuterations can be followed by infrared and n.m.r. spectroscopy. That the terminal hydrogen atoms remain unaffected in the early stages of the deuteration is shown by the persistence of their characteristic absorptions in the infrared and n.m.r. spectra of the products, whereas the absorptions due to the bridging hydrogens shift to lower frequency in the infrared spectrum and disappear from the n.m.r. spectra.

Salts of the anion $B_{10}H_{13}^-$ are accessible by the action of alkali metal hydrides or tetrahydroborates on decaborane (an excess of sodium hydride gives the anion $B_{10}H_{12}^{2-}$—see Fig. 4–15c and d), or by abstraction of a proton using a suitable base [for example, certain amines, triphenyl-(methylene)phosphorane]. These salts tend to contain variable proportions of ether solvent unless the cation is very large, a feature taken to indicate that metal–boron interactions with the anion $B_{10}H_{13}^-$ are strong in the case of small cations. Consistent with this, the major product of the reaction between decaborane and the methyl Grignard reagent MeMgI (a minor product is $6\text{-MeB}_{10}H_{13}$) is the compound $B_{10}H_{13}MgI$ which has properties more like those of a Grignard reagent RMgX than those of a salt $Mg^{2+}I^-B_{10}H_{13}^-$, and is accordingly frequently referred to as the decaboranyl Grignard reagent: it is a useful intermediate from which to prepare 6-substituted decaboranes.

$$B_{10}H_{14} \xrightarrow{\text{MeMgI}} B_{10}H_{13}MgI \xrightarrow{\text{RX}} 6\text{-RB}_{10}H_{13}$$

(for example, RX = $PhCH_2Cl$, AlkF, Alk_2SO_4, Ph_2PCl)

The anion $B_{10}H_{13}^-$ is an intermediate in the alkaline degradation of decaborane(14) to nonaboranes. Aqueous hydroxide ion degrades decaborane apparently by the following steps:

$$B_{10}H_{14} \xrightarrow{\text{OH}^-} B_{10}H_{13}^- \xrightarrow{\text{OH}^-} B_{10}H_{13}OH^{2-} \xrightarrow{\text{H}_3\text{O}^+} B_9H_{14}^- + B(OH)_3 + H_2$$

A topological representation of the product is in Fig. 4–15h. Note that the boron atom removed is boron-9, i.e. one of the two borons of decaborane(14) most susceptible to nucleophilic attack, and presumably that still susceptible to nucleophilic attack in the anion $B_{10}H_{13}^-$ (see Fig. 4–15b).

On the basis of his topological theory, Lipscomb predicted that decaborane(14) should be capable of reduction to the anion $B_{10}H_{14}^{2-}$, for which he predicted the structure shown in Fig. 4–15e, with two bridging hydrogens linking the 5,10 and 7,8 boron pairs, and borons 6 and 9 each bearing two terminal hydrogens. This anion has since been prepared by the following routes:

$$B_{10}H_{14} + 2Na \xrightarrow{\text{liquid } NH_3 \text{ or ether solvent}} 2Na^+ B_{10}H_{14}^{2-}$$

$$B_{10}H_{14} \xrightarrow{-H^+} B_{10}H_{13}^- \xrightarrow{BH_4^-} B_{10}H_{14}^{2-} + B(OH)_3 + 3H_2$$

The structure (Fig. 4–15e) has been confirmed by X-ray crystallography for some isoelectronic species $B_{10}H_{12}L_2$ (L, an electron pair donor, and H^- are isoelectronic substituents in these compounds). Acidification of aqueous solutions of the anion $B_{10}H_{14}^{2-}$ generates the anion $B_{10}H_{15}^-$, alternative plausible structures for which are shown in Fig. 4–15f and g.

A particularly interesting series of compounds which has been quite extensively studied is that series formally related to the anions $B_{10}H_{13}^-$, $B_{10}H_{14}^{2-}$, $B_{10}H_{15}^-$, and $B_9H_{14}^-$ but with a Lewis base species L replacing a hydride ion H^- on either or both of boron atoms 6 and 9. Their relationships to the hydride anions and to each other are illustrated in Table 4–9 and Figs 4–16 and 4–18. It is seen that Lewis bases can react with decaborane(14) in two distinct ways, the first of which (abstraction of a bridging hydrogen as a proton) has already been mentioned as the preferred rapid interaction with water. Alternatively and more slowly in the case of water, a Lewis base may become attached to one of the two boron atoms most susceptible to nucleophilic attack, i.e. borons 6 or 9. Once the first Lewis base species has become attached to the 6-position to form the 1:1 adduct $B_{10}H_{14}$,L, elimination of hydrogen and attachment of a second Lewis base species at the 9-position occurs rapidly, so that the normal product of reactions between decaborane(14) and a wide range of Lewis bases L is the derivative $B_{10}H_{12}L_2$.

$$B_{10}H_{14} \xrightarrow[\text{slow}]{L} 6\text{-}B_{10}H_{14},L \xrightarrow[\text{fast}]{-H_2} B_{10}H_{12},L \underset{-L}{\overset{+L}{\rightleftharpoons}} 6,9\text{-}B_{10}H_{12}L_2$$

The reversibility of the last stage of this reaction sequence has allowed the relative donor properties of different bases L to the $B_{10}H_{12}$ residue to be compared by displacement reactions:

$$B_{10}H_{12}L_2^1 + 2L^2 \rightleftharpoons B_{10}H_{12}L^1L^2 + L^1 + L^2 \rightleftharpoons B_{10}H_{12}L_2^2 + 2L^1$$

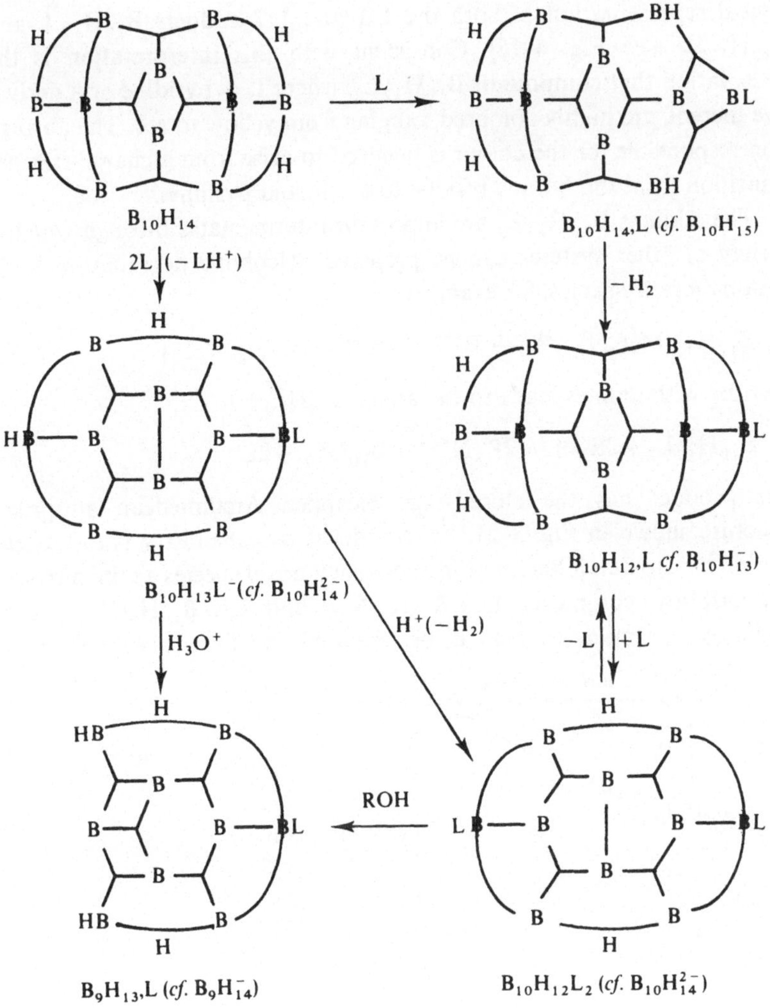

In all these structures, B represents the BH group.

Fig. 4-18 Bonding representations of some species resulting from reactions between decaborane(14) and Lewis bases.

These have shown donor properties to increase in the sequence:

$$R_2S \sim RCN < R_2NCN < R_2NCOR < R_3N < R_3P \sim pyridine$$

Lewis bases to the right displace Lewis bases to the left. Note that such π-acid ligands as phosphines and pyridine coordinate more strongly than do amines, a feature interpreted as indicating that back-bonding from the boron framework contributes to the strength of binding of L. Such back-bonding is readily intelligible in that the boron to which L is attached is the central boron of an open BBB bond in the simplest topological representation of both the 1:1 and 1:2 adducts $B_{10}H_{12}$,L and $B_{10}H_{12}L_2$ (see Fig. 4–18). Consistent with this interpretation is the observation that compounds $B_{10}H_{12}L_2$, where L = pyridine or a derivative thereof, are highly coloured, ranging from yellow to red. The absorption responsible for the colour is believed to arise from a charge-transfer transition from the borane residue to the ligand pyridine.

The adducts $B_{10}H_{12}L_2$ are important intermediates through which a variety of other systems can be prepared. Alcoholic degradation leads to nonaborane species, for example,

$$B_{10}H_{12}(MeCN)_2 \xrightarrow{MeOH} B_9H_{13},MeCN \text{ (see Fig. 4–18)}$$

Heating with amines leads to the anion $B_{10}H_{10}^{2-}$:

$$B_{10}H_{12}L_2 + 2R_3N \rightarrow 2R_3NH^+ + B_{10}H_{10}^{2-} + 2L$$

The product has the closed-cage bicapped Archimedean antiprism structure shown in Fig. 3–21. Its own derivative chemistry is too extensive[12,24–26] to detail here, but includes such novel species as the nitrogen and carbonyl complexes $1,10\text{-}B_{10}H_8(N_2)_2$ and $1,10\text{-}B_{10}H_8(CO)_2$ from which in turn many others can be prepared (see Fig. 4–19). The icosa-

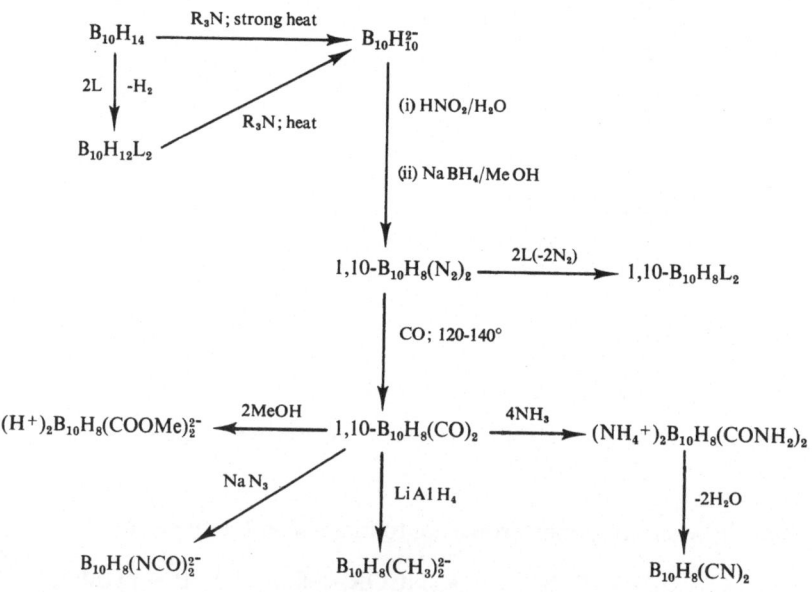

Fig. 4–19　Some derivatives of the anion $B_{10}H_{10}^{2-}$

hedral anion $B_{12}H_{12}^{2-}$, accessible through high temperature reactions between tetrahydroborates MBH_4 or amine-boranes R_3N,BH_3 and diborane, has a similarly vast derivative chemistry (see, for example, refs 12, 24, 25, and 26).

The adducts $B_{10}H_{12}L_2$ are intermediates in the preparation of yet another important class of closed-cage system, the carboranes. Icosahedral carboranes $R_2C_2B_{10}H_{10}$ result from reactions with acetylenes:

$$B_{10}H_{14} \xrightarrow[-H_2]{2L} B_{10}H_{12}L_2 \xrightarrow{RC\vdots CR} R_2C_2B_{10}H_{10}$$

The chemistry of carboranes is outlined in chapter 6.

Further examples of reactions between Lewis bases and decaborane(14) or its derivatives are in Table 4–10. Note that, as with diborane, decaborane(14) accommodates ligands by rearrangements in which the numbers of bridging hydrogen atoms are reduced. The molecule cannot accommodate an extra pair of bonding electrons without structural modification.

References

1. Wagman, D. D., W. H. Evans, V. B. Parker, I. Halow, S. M. Bailey, and R. H. Schumm, Selected Values of Chemical Thermodynamic Properties, National Bureau of Standards Technical Note 270–3, January, 1968.
2. Stock, A., Hydrides of Boron and Silicon, Cornell University Press, Ithaca, New York, 1933.
3. See, for example, Sanderson, R. T., Vacuum Manipulation of Volatile Compounds, Wiley, New York, 1948; Shriver, D. F., The Manipulation of Air-sensitive Compounds, McGraw-Hill, New York, 1969.
4. Schlesinger, H. I., H. C. Brown, and collaborators, J. Amer. Chem. Soc., 1953, 75, 186.
5. Holzmann, R. T. (ed.), R. L. Hughes, I. C. Smith, and E. W. Lawless, Production of the Boranes and Related Research, Academic Press, New York, 1967.
6. Long, L. H., J. Inorg. Nucl. Chem., 1970, 32, 1097.
7. Stone, F. G. A., Chemical reactivity of the boron hydrides and related compounds, Advances in Inorg. Chem. and Radiochem., 1960, 2, 279–314.
8. Schenker, E., Uses of complex borohydrides and of diborane in organic chemistry, Angew. Chem., 1961, 73, 81.
9. Lipscomb, W. N., Boron Hydrides, W. A. Benjamin, New York, 1963.
10. Adams, R. M., The Boranes or Boron Hydrides, pp. 507–692 of ref. 11.
11. Adams, R. M. (ed.), Boron, Metallo–boron Compounds, and Boranes, Interscience, New York, 1964.
12. Hawthorne, M. F., Boron Hydrides, pp. 223–324 of The Chemistry of Boron and its Compounds, E. L. Muetterties (ed.), Wiley, New York, 1967.
13. Brown, H. C., Hydroboration, W. A. Benjamin, New York, 1962.
14. Shore, S. G., and C. L. Hall, J. Amer. Chem. Soc., 1967, 89, 3947.
15. Burg, A. B., J. Chem. Educ., 1960, 37, 482.
16. Purcell, K. F., J. Amer. Chem. Soc., 1969, 91, 3487; Lloyd, D. R., and N. Lynaugh, Chem. Comm., 1970, 1545.
17. Brown, M. P., R. W. Heseltine, P. A. Smith, and P. J. Walker, J. Chem. Soc. (A), 1970, 410.
18. Finn, P. A., and W. L. Jolly, Chem. Comm., 1970, 1090.
19. Cotton, F. A., and J. R. Leto, J. Chem. Phys., 1959, 30, 993.
20. Gaines, D. F., and R. Schaeffer, J. Amer. Chem. Soc., 1964, 86, 1505.
21. Brown, D. R., S.F.A. Kettle, J. McKenna, and J. M. McKenna, Chem. Comm., 1967, 667.
22. Odom, J. D., and R. Schaeffer, Use of Isotopic Labels in the Study of Carboranes and Binary Compounds of Boron and Hydrogen, pp. 141–172 of ref. 23.

23. Brotherton, R. J., and H. Steinberg (eds.), *Progress in Boron Chemistry*, vol. 2. Pergamon Press, Oxford, 1970.
24. Adams, R. M., The Hydroboron Ions (Ionic Boron Hydrides), pp. 373–506 of ref. 11.
25. Muetterties, E. L., and W. H. Knoth, *Polyhedral Boranes*, Marcel Dekker, New York, 1968.
26. Todd, L. J., The Chemistry of Polyhedral Borane Ions, pp. 1–36 of ref. 23.
27. Hawthorne, M. F., Decaborane(14) and its derivatives, *Advances in Inorg. Chem. and Radiochem.*, 1963, **5**, 307.

Problems

4–1   Under suitable conditions, most boron hydrides can be hydrolysed quantitatively to boric acid and hydrogen, according to the equation:

$$B_pH_{p+q} + xH_2O \rightarrow yB(OH)_3 + zH_2$$

(a) Express x, y, and z as functions of p and/or q.

(b) Calculate the enthalpy change ΔH (hydrolysis) for the gas-phase hydrolysis of the hydrides B_2H_6, B_4H_{10}, B_5H_9, B_5H_{11}, and B_6H_{10}, using the data in Table 4–1, and values of 2960 kJ/mole and 927 kJ/mole for the gas-phase heats of atomization of boric acid and water respectively. Express your results (i) in kJ/g mole and (ii) in kJ/g, and comment on the extent to which the latter vary from one hydride to another.

4–2   The amounts of heat evolved (ΔH kJ/mole) when adducts L,BH_3 are formed in the gas-phase from diborane and a Lewis base L according to the equation:

$$\tfrac{1}{2}B_2H_6 + L \rightarrow L,BH_3 + \Delta H$$

are as follows:

L	Me_3N	Me_3As	Me_2S	Me_2Se	CO
ΔH	42	46	45	34	20

Calculate values of the ligand $\rightarrow$ boron dative bond energy term for these complexes, stating any assumptions made, and comment on the relative capacities of these donors to borane(3). Suggest a likely minimal value for ΔH for $L = Me_3P$.

4–3   Values of the bond energy terms E(B–X) (in kJ/mole) for the boron trihalides BX_3, E(X–X) for the halogens X_2, and E(H–X) for the hydrogen halides HX are given below. Using them, and a value of 436 kJ/mole for E(H–H) for the hydrogen molecule, calculate the enthalpy changes for the following reactions (X = F, Cl, Br, or I):

(a) $B_2H_6 + 6HX \rightarrow 2BX_3 + 6H_2$

(b) $B_2H_6 + 6X_2 \rightarrow 2BX_3 + 6HX$

Comment on the chemical significance of the variations in enthalpy with X.

	F	Cl	Br	I
E(B–X)	644	444	368	272
E(X–X)	158	244	193	151
E(H–X)	568	432	366	298

4–4   Explain how one might separate in the vacuum line the components of mixtures made from the following substances:

(a) hydrogen, diborane, and tetraborane(10);

(b) diborane, pentaborane(9), and decaborane(14);

(c) an equimolar mixture of diborane and trimethylamine.

4–5   Survey those reactions in which boron hydrides function as Lewis acids.

4-6 Discuss the extent to which the reactions of diborane and the higher boron hydrides reflect the electron deficiencies of their molecular skeletons.

4-7 Suggest alternative bonding representations for the anions shown in Fig. 4-15, retaining essentially the same arrangements of the boron atoms, but changing the numbers of bridging hydrogens and BH_2 groups if appropriate.

4-8 Write down the formulae of compounds which might in principle be prepared by reactions between diborane and dimethylamine, taken in varying proportions, and with or without hydrogen elimination. Where possible, indicate what reaction conditions (proportions of reactants, temperature) would be appropriate for their synthesis, and what simple chemical, physical, or spectroscopic tests (apart from elemental analyses) could be used for their identification.

4-9 Discuss the extent to which the reactions of decaborane(14) reflect the charge distribution in the molecule shown in Fig. 4-17.

4-10 Give the conditions for, and products expected from, reactions between diborane and the following:
(a) ethanol, (b) sodium hydride, (c) propene, (d) but-2-yne, (e) acetone, (f) trimethylborane, (g) trimethylaluminium, (h) trimethylethylene, (i) molecular deuterium, (j) acetonitrile.

4-11 Discuss the relative tendencies of three-centre BHB bonds and three-centre BBB bonds to function as sources of reactivity in the boron hydrides.

4-12 Describe the effect of the following reagents on (i) diborane, (ii) pentaborane(9), and (iii) decaborane(14): (a) water, (b) sodium hydride, (c) trimethylamine.

4-13 Outline a method by which a sample of each of the following compounds might be prepared, using readily available sources of deuterium: (a) fully deuterated diborane, $B_2{}^2H_6$; (b) fully deuterated hexaborane(10), $B_6{}^2H_{10}$; (c) fully deuterated pentaborane(9), $B_5{}^2H_9$; (d) pentaborane(9) deuterated in all the terminal, but not the bridging, positions.

4-14 Outline a method by which a sample of each of the following compounds might be prepared:
(a) chlorodiborane, B_2H_5Cl; (b) 1-methylpentaborane(9); (c) 2-methylpentaborane(9); (d) 2-ethyldecaborane(14); (e) disodium dodecahydrodecaborate(2−); (f) disodium decahydrodecaborate(2−).

4-15 Reaction of diborane with an equimolar proportion of 1,3-butadiene gave a product, compound **A**, $C_4H_{12}B_2$, the ^{11}B n.m.r. spectrum of which consisted of a doublet of triplets. The infrared spectrum of **A** contained a single absorption near 2500 cm^{-1}, and other absorptions in the region 1500–2100 cm^{-1}. On hydrolysis, 20·5 mg of **A** evolved one mmole of hydrogen, leaving a residual acid $C_4H_{12}B_2O_4$. When treated with an excess of trimethylamine, **A** gave an adduct, **B**, which had a molecular weight of *ca.* 200. The infrared spectrum of **B** contained two absorptions near 2500 cm^{-1}, but no major absorption in the range 1500–2100 cm^{-1}, and its ^{11}B n.m.r. spectrum consisted of a 1:2:1 triplet.

Identify **A** and **B**, explaining the steps in your reasoning and the significance of the various pieces of information given.

4-16 A mixture of boron hydrides was separated into its components in the vacuum line, using a series of U traps cooled to −78°, −95°, and −196°C. The compound (**A**) that collected in the first trap had a vapour density of 3·34 mg/cc (corrected to s.t.p.), and generated 14 times its own volume of hydrogen when hydrolysed by acid. Compound **B** which collected in the trap at −95° had an ^{11}B n.m.r. spectrum which superficially resembled that of **A**, consisting of two doublets, though the relative intensities of these doublets differed in the two spectra. Treatment of 21 mg of **B** with an excess of alcohol slowly generated 89·6 cc of hydrogen (corrected to s.t.p.). Compound **C**, which collected in the trap at −196°C, generated 11 times its own volume of hydrogen when shaken with hot water, and reacted with an equimolar proportion of ethylene to eliminate a mole of hydrogen per mole of ethylene consumed.

Identify **A**, **B**, and **C**, explaining the steps in your reasoning and the significance of the various pieces of information given.

4–17 Trimethylborane (44·8 cc) and diborane (44·8 cc) were mixed and reacted in the gas phase to form compound **A** as the major product. Methyl cyanide (0·246 g) was added, and the mixture was left at 10° for 5 days. Subsequent fractionation in the vacuum line afforded compounds **B**, **C**, and **D**.

B was a colourless liquid which decomposed when exposed to moist air. With an excess of alkali, 0·055 g of **B** gave hydrogen (22·4 cc), ethylamine (1 mmole), and a solution containing borate ions. The parent peak in its mass spectrum was at $m/e = 165$. The main features of its ^1H n.m.r. spectrum were a quartet at 6·47 τ, relative intensity 2, and a triplet at 8·64 τ, relative intensity 3 ($\tau(Me_4Si) = 10\cdot00$ ppm).

C was a colourless crystalline solid of molecular formula $C_8H_{20}B_2N_2$. Not air sensitive, it was hydrolysed above 100°C by concentrated aqueous alkali, forming ammonia (1 mmole per 0·083 g of **C**) but no hydrogen. Its ^1H n.m.r. spectrum contained the following peaks (relative intensities in parentheses): a quartet at 2·64 τ (1), a doublet at 8·03 τ (3), and a singlet at 10·14 τ (6).

D was a colourless liquid isomer of **C**, with similar chemical reactions and ^1H n.m.r. spectrum, except that the peak at 10·14 τ was replaced by two peaks, each of relative intensity 3, at 10·00 and 10·21 τ.

Identify **A**, **B**, **C**, and **D**, explaining the steps in your reasoning. What would be the effect of (i) water and (ii) trimethylamine on **A**? (Gas volumes are at s.t.p.; 1 mmole of gas occupies 22·4 cc.)

5 Associated metal hydrides, alkyls, and aryls

5–1   Introduction

In previous chapters, the three-centre two-electron hydrogen bridge has been shown to be a common structural feature of all the neutral boron hydrides and of many of their derivatives, and a source of many of their characteristic reactions, particularly those in which the bridge is cleaved by nucleophiles (Lewis bases). Similar electron deficient hydrogen bridges linking atoms of many other elements than boron are now known, and it is the object of the present chapter to indicate briefly in what types of compound they occur. These types are too numerous to allow also detailed discussion of their reactions, many of which can, however, be inferred by analogy with those of the boron hydrides described in Chapter 4.

Electron deficient hydrogen bridges occur in the following classes of compound:

1. In binary hydrides MH_x of elements to the left of or below boron in the periodic table (see Fig. 5–1). For example, in aluminium hydride, $(AlH_3)_n$, all the hydrogen atoms are bridging hydrogens (see Fig. 5–2).

2. In compounds derived from these binary hydrides by replacement of some of their hydrogen atoms by σ-bonded alkyl or aryl groups, as in dialkyl-aluminium or -gallium hydrides $(R_2MH)_{2 \text{ or } 3}$.

3. In certain mixed hydrides, which contain hydrogen bridges between one element and another, for example, between a transition or main group metal and boron in certain tetrahydroborates $M(BH_4)_x$, (Figs 5–3 → 5–7), or between lithium and aluminium in lithium tetrahydroaluminate, $LiAlH_4$.

4. In various hydride complexes, particularly polynuclear transition metal carbonyl hydrides $H_xM_y(CO)_z$, exemplified by the anionic chromium, molybdenum and tungsten species $[HM_2(CO)_{10}]^-$, and the neutral ruthenium complex $H_2Ru_6(CO)_{18}$ (Fig. 5–8).

Although it is particularly difficult to locate the hydrogen atoms unambiguously in many of these systems, it is already apparent that a variety of coordination numbers and environments is possible for the bridging hydrogen. Some indication of this variety is given below.

Also included in this chapter is a brief survey of the types of structure possible when carbon, rather than hydrogen, occupies an electron deficient bridging position. Carbon displays a versatility comparable to that of hydrogen in forming electron deficient bridges. Alkyl, alkenyl, alkynyl, and aryl groups attached to elements more electropositive than boron can act as bridges between two or more main group metal atoms, and carbon atoms with coordination numbers greater than four have been found in polynuclear transition metal complexes like the ruthenium carbonyl complex $Ru_6C(CO)_{17}$, in which the central carbon is surrounded octahedrally by metal atoms.

5–2   Associated metal hydrides

Binary hydrides MH_x may be classified in a manner that reflects the position of the element M in the periodic table (see Fig. 5–1). Boron occupies a key position in such a classification. Elements to its right in the

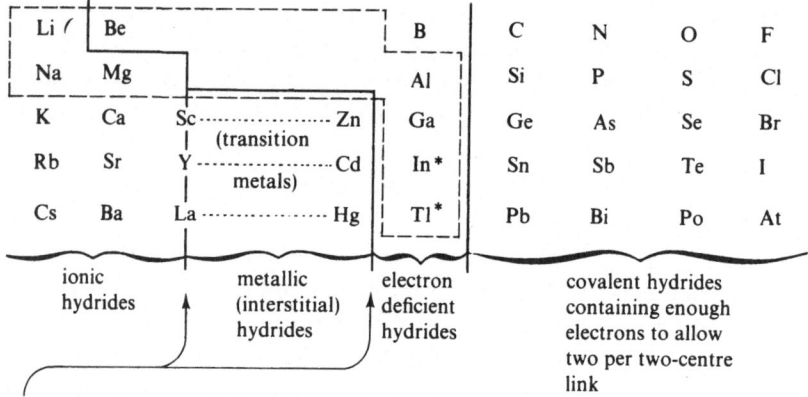

The ionic/metallic/electron deficient boundaries are not clear-cut
Compounds on either side may be regarded as borderline species

The elements enclosed by the broken line form electron deficient bridged alkyls and/or aryls

* The hydrides of these elements are of low thermal stability, have been little studied, and are of uncertain type

Fig. 5–1   Classification of binary hydrides MH_x

periodic table (i.e. elements of the p-block groups 4, 5, 6, and 7) form covalent hydrides the molecules of which have enough valence electrons to allocate two per M–H bond. On the other hand, elements to the left of or below boron in the periodic table form hydrides in which, as in the boron hydrides, the number of near-neighbour contacts between hydrogen atoms and atoms of the element M exceeds the number of available pairs of bonding electrons, and in that sense these hydrides may all be considered electron deficient. However, the hydrides of the most electropositive metals are best treated as ionic, with cations M^{x+} and anions H^- packed in lattices appropriate for their relative numbers and sizes.

Moreover, many of the transition elements known to form binary hydrides do so by incorporating hydrogen atoms in suitable interstices (frequently tetrahedral holes) in essentially metallic lattices, so that a convenient treatment of their bonding is to regard it as metallic, with both metal and hydrogen atoms contributing electrons to incompletely filled energy bands.

There remain the hydrides of elements immediately to the left of or below boron in the periodic table, i.e. beryllium, aluminium, and gallium. The bonding in these hydrides may be expected to have a high degree of covalent character, and the structural information available indicates that they contain MHM hydrogen bridges. They are exceedingly difficult to prepare in a pure and suitably crystalline form for single crystal X-ray diffraction work. However, the structure of a hexagonal modification of aluminium hydride $(AlH_3)_n$ (several other crystalline forms are known) has been elucidated from X-ray powder diffraction data and from neutron-diffraction data for the deuterated material $(Al^2H_3)_n$.[2] The environment of each aluminium atom in the polymeric lattice is as shown in Fig. 5–2.

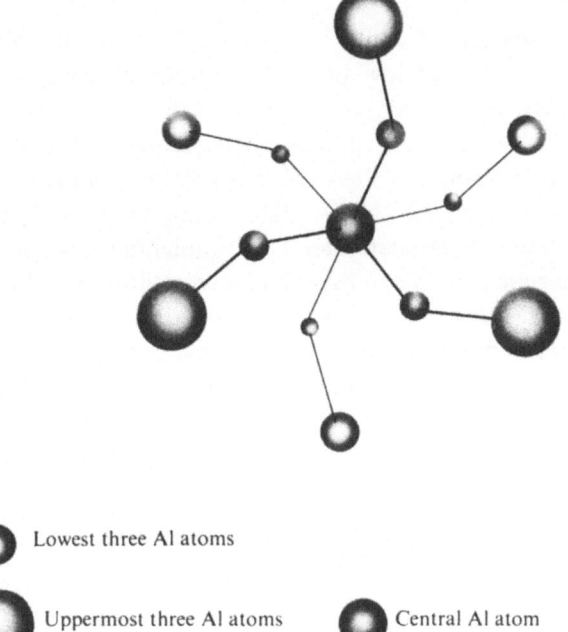

Lowest three Al atoms

Uppermost three Al atoms   Central Al atom

Lower three H atoms   Upper three H atoms

Interatomic distances: Al . . . . H, 172 pm; Al . . . . Al, 324 pm

Fig. 5–2   The six AlHAl bonds emanating from a single aluminium atom in hexagonal aluminium hydride

Ⅰ

Each aluminium atom is surrounded octahedrally by six bridging hydrogen atoms, each of which bridges it to a different neighbouring aluminium atom. The involvement of each atom in *six* three-centre two-electron bonds reflects the ability of aluminium to use more orbitals for bonding (a 3*s*, three 3*p*, and two 3*d* orbitals) than can boron (restricted to a 2*s* and three 2*p* orbitals). Since the bridges in aluminium hydride are AlHAl bridges, not AlH$_2$Al bridges—pairs of aluminium atoms are linked by only *one* hydrogen atom—the AlHAl bond angle is much larger than the BHB bond angle in diborane. The spherical symmetry of the hydrogen 1*s* orbital allows it to overlap with the metal orbitals to an extent which is independent of the MHM bond angle.

The method of preparation of the unsolvated aluminium hydride sample used for the above structural study was not reported, presumably because of the interest in such substances as solid rocket fuels. When prepared by the addition of aluminium chloride to an ethereal solution of lithium tetrahydroaluminate, aluminium hydride is difficult to separate from coordinated solvent:

$$AlCl_3 + 3LiAlH_4 \xrightarrow{Et_2O} 3LiCl + 4/m\,(AlH_3)_m(Et_2O)_n$$

The evaporation of aluminium from a hot tungsten filament in a stream of hydrogen at low pressure leads to monomeric aluminium hydride, AlH_3, in the gas phase, which if its concentration builds up dimerizes to Al_2H_6, as shown by mass spectroscopic study of the gas stream just beyond the filament.[3] Peaks of mass/charge ratio $m/e = 30$ and 60 arise due to the ions AlH_3^+ and $Al_2H_6^+$, shifting to $m/e = 33$ and 66 if the hydrogen is replaced by deuterium. These aluminium analogues of borane and diborane are expected to have similar structures to the boron species. They associate further if trapped by condensation on a cold surface, otherwise decompose again to aluminium and hydrogen.

A derivative of dialane, μ-dimethylaminodialane, $Me_2NAl_2H_5$, has been prepared from an aluminium hydride solution in diethyl ether:[4]

$$2AlH_3 + Me_2NH \xrightarrow{Et_2O} H_2 + H_2Al \underset{\underset{Me_2}{N}}{\overset{H}{\diamond}} AlH_2$$

Uncoordinated gallium hydride $(GaH_3)_n$ can be obtained as a viscous liquid, f.p. $-15°C$, by the following route:[5]

$$GaCl_3 \xrightarrow[-3LiCl]{4LiH} LiGaH_4 \xrightarrow[-LiCl,-H_2]{Me_3NHCl} Me_3N,GaH_3 \xrightarrow[-Me_3N,BF_3]{BF_3} 1/n(GaH_3)_n$$

Early reports that the product is dimeric (Ga_2H_6) have not been substantiated by subsequent work. Its involatility, viscosity, and immiscibility with non-polar solvents are consistent with a high degree of association, though its thermal instability (it decomposes to gallium and hydrogen at room temperature) and insolubility in suitable solvents has prevented confirmation of this by molecular weight determination. Evaporation of gallium in a hydrogen stream affords the monomer GaH_3 (detected by mass spectroscopy) which, however, has too low thermal stability to build up sufficient pressure to generate dimers Ga_2H_6. The trimethylamine adduct Me_3N,GaH_3 contains tetrahedrally coordinated gallium, and only terminal Ga–H bonds (by X-ray crystallography). Its Ga–H stretching absorption in the infrared spectrum occurs at 1853 cm^{-1}. The corresponding absorption in the spectrum of uncoordinated gallium hydride is at *higher* frequency (1980 cm^{-1}) indicating that the latter also probably contains terminal Ga–H bonds, and is therefore less extensively bridged than is aluminium hydride, as indeed the low melting point also appears to show.

Beryllium hydride $(BeH_2)_n$ is accessible in varying degrees of purity by such reactions as the following:[6]

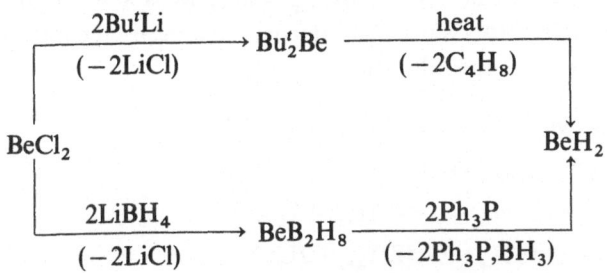

The product tends to be contaminated by the materials from which it has been prepared, as these get trapped in the polymeric lattice. It is believed to contain four-coordinate beryllium atoms and two-coordinate hydrogens (the metal:hydrogen ratio requires the hydrogen coordination number to be half that of the metal). The hydrogen atoms probably bridge singly (BeHBe) rather than in pairs (BeH_2Be) between adjacent metal atoms (cf. aluminium hydride).

Beryllium's more electropositive neighbours, lithium and magnesium, form hydrides which may be regarded as borderline between ionic and covalent. Lithium hydride, like the other alkali metal hydrides, has a rock-salt lattice in which both metal and hydrogen are six-coordinate, while magnesium hydride has the rutile structure, in which the metal is six-coordinate and hydrogen three-coordinate. In both these hydrides, also in calcium, strontium, and barium hydrides and to a small extent in sodium hydride, polarization of the anion by the small cations makes its effective radius smaller than that in the heavier alkali metal hydrides (see Table 5–1).[1]

Table 5–1   Effective radius of the hydride anion, $r(H^-)$, in Group I and Group II metal hydrides

MH:	M	Li	Na	K	Rb	Cs
	M–H distance (pm)	204	245	286	303	320
	$r(H^-)$(pm)	137	146	152	154	152
MH$_2$:	M	Be	Mg	Ca	Sr	Ba
	M–H distance (pm)	—	195	232	249	267
	$r(H^-)$(pm)	—	131	133	136	132

5–3   Some mixed hydrides

Certain mixed hydrides, particularly the covalent tetrahydroborates $M(BH_4)_x$,[7] provide better examples than do the simple binary hydrides MH_x of the capacity of a wide range of metals to participate in three-centre two-electron hydrogen bridge bonds.

Many tetrahydroborates $M(BH_4)_x$, where M is a single atom of an element more electropositive than boron, have been prepared by means of such reactions as the following:

$$4NaH + (MeO)_3B \rightarrow NaBH_4 + 3NaOMe$$
$$2LiH + B_2H_6 \rightarrow 2LiBH_4$$
$$(Me_3Al)_2 + 4B_2H_6 \rightarrow 2Al(BH_4)_3 + 2Me_3B \cdot$$
$$BeCl_2 + 2LiBH_4 \rightarrow Be(BH_4)_2 + 2LiCl$$

Attempts at the preparation of tetrahydroborates of more electronegative elements lead to the hydride and diborane:

$$SiCl_4 + 4LiBH_4 \rightarrow SiH_4 + 2B_2H_6 + 2LiCl$$

Tetrahydroborates have properties that reflect the polarizing power and electropositive character of M. Derivatives of the alkali metals are essentially ionic $M^+BH_4^-$, and are involatile air- and moisture-stable crystalline solids. Neutron-diffraction studies on the potassium salt showed the tetrahedral anion to have a B–H bond length of 126 pm. Less electropositive metals that form small or highly charged cations form volatile, highly reactive covalent tetrahydroborates in which the metal is bridged to boron through two or even three of the hydrogens of the tetra-hydroborate residue. Examples are the aluminium compound, $Al(BH_4)_3$ (Fig. 5–3), and the zirconium compound, $Zr(BH_4)_4$ (Fig. 5–4).

In the aluminium compound, the structure of which was determined by electron-diffraction using a gaseous sample, the six bridging hydrogens

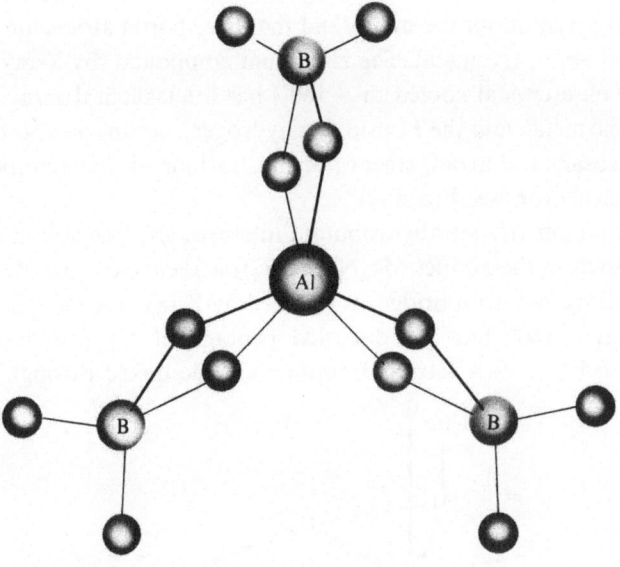

Interatomic distances (pm):
Al–B, 214; Al–H$_\mu$, 180; B–H$_\mu$, 128; B–H$_t$, 120

Bond angles:
∠ H$_\mu$AlH$_\mu$, 73°   ∠ H$_\mu$BH$_\mu$, 114°   ∠ H$_t$BH$_t$, 116°

Fig. 5–3   The molecular structure of aluminium tris (tetrahydroborate), Al(BH$_4$)$_3$

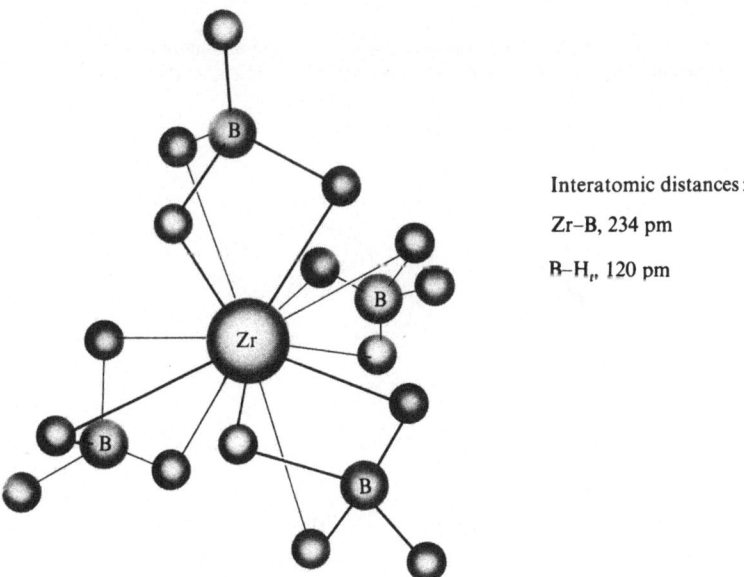

Interatomic distances:

Zr–B, 234 pm

B–H$_t$, 120 pm

Fig. 5–4   Molecular structure of zirconium tetrakishydroborate, Zr(BH$_4$)$_4$, showing the triple hydrogen bridges between zirconium and boron

probably (there is slight uncertainty about their precise positions) form a trigonal prism about the metal, and the three boron atoms are trigonally disposed about the metal. The zirconium compound (by X-ray crystallo-graphy on a crystal cooled to $-160°$) has a tetrahedral array of borons about the metal, and the 12 bridging hydrogens occupy a cubo-octahedral or an icosahedral arrangement (for illustrations of the cubo-octahedron and icosahedron, see Fig. 6–4).

Aluminium tris-tetrahydroborate interestingly reacts with trimethyl-amine to form the adduct $Me_3N,Al(BH_4)_3$ *without* cleavage of an alumin-ium–hydrogen–boron bridge, as revealed by X-ray studies of the product, which at $-160°$ has the distorted pentagonal bipyramidal structure shown in Fig. 5–5. A tetrahydroborate residue linked through two of its

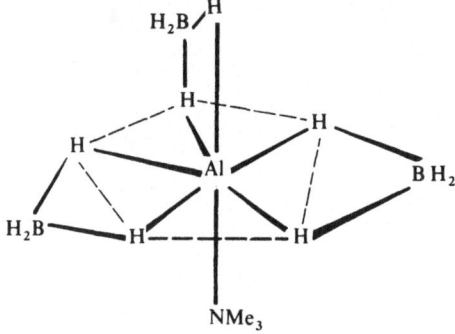

Interatomic distances (pm; $-160°$):
Al–N, 200; Al–B, 223; Al–H$_\mu$, 197; B–H$_\mu$, 143; B–H$_t$, 119

Fig. 5–5 Pentagonal bipyramidal coordination about aluminium in the aluminium tris-hydroborate–trimethylamine adduct $Me_3N,Al(BH_4)_3$

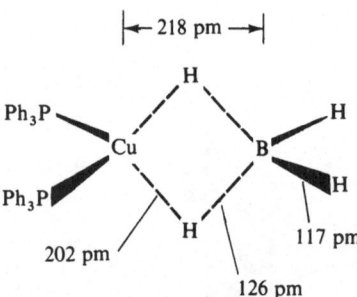

Fig. 5–6 Molecular structure of bis(triphenylphosphine)copper(I) tetrahydroborate $(Ph_3P)_2CuBH_4$

hydrogens to the metal is also present in the copper(I) complex $(Ph_3P)_2$-$CuBH_4$ (Fig. 5–6), though not in the beryllium compound BeB_2H_8, for which the bis-tetrahydroborate structure $H_2BH_2BeH_2BH_2$ (Fig. 5–7a) was long assumed.

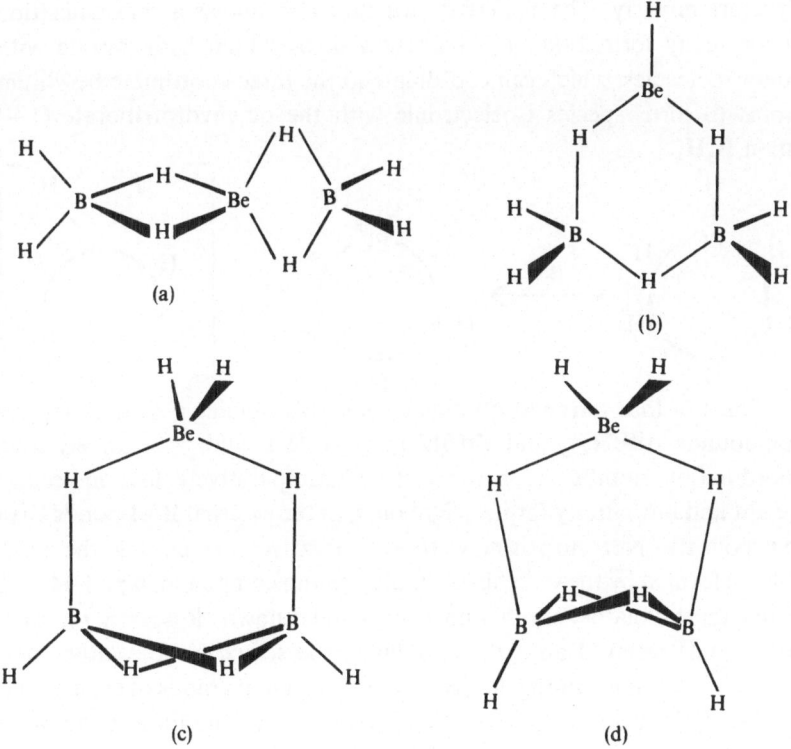

Fig. 5–7 Structures considered for beryllium diboron octahydride, BeB_2H_8

Beryllium diboron octahydride BeB_2H_8 is a volatile highly reactive air- and moisture-sensitive solid which has presented a fascinating structural problem. Its structure has not been solved by X-ray crystallography—it contains only light atoms and crystallizes most inconsiderately with 16 molecules in the unit cell of its tetragonal lattice. Monomeric in the gas phase, it diffracts electrons in a manner appropriate for a triangular, rather than linear, arrangement of the three 'heavy' atoms, the beryllium and two borons, as in Fig. 5–7b, c, or d. Although all three are consistent with the electron-diffraction data, the first (Fig. 5–7b) appears to give the most reasonable set of interatomic distances, and moreover its terminal BH_2 (as opposed to BH) groups are indicated by such infrared spectroscopic features as the doublet absorption (at 2630 and 2531 cm^{-1}, shifting to 1910 and 1816 cm^{-1} for the deuterated compound $BeB_2{}^2H_8$) attributable to terminal B–H stretching vibrations.[8] The fragmentation of the compound in the mass spectrometer shows the beryllium atom to be strongly bound to the rest of the molecule, and the dipole moment of $2\cdot1\pm0\cdot5D$ at $25°$ rules out the linear structure (Fig. 5–7a)—there are too many bands in the infrared spectrum for the linear structure anyway. The preferred structure also allows a rationalization of the ready formation of 1:1 adducts of beryllium hydroborate with donor molecules, which can coordinate to the three-coordinate beryllium atoms to form species isoelectronic with the octahydrotriborate $(1-)$ anion $B_3H_8^-$:

Three or four tetrahydroborate groups surrounding a metal, as in the compounds $Al(BH_4)_3$ and $Zr(BH_4)_4$, provide it with a relatively high coordination number in a molecule of comparatively low molecular weight and low affinity for its neighbours, as the external B–H bonds have low polarity. Not surprisingly, these derivatives are among the most volatile formed by these metals—the aluminium compound, b.p. $+44.5°C$, is the most volatile aluminium compound known. Research on such compounds received an early stimulus in the search for a volatile compound of uranium suitable for gas-phase diffusion methods of enrichment of the ^{235}U isotope for atomic weapons. The uranium compound $U(BH_4)_4$, can be prepared from the tetrafluoride UF_4 and aluminium hydroborate; it is a green solid, volatile above 30°, with an infrared spectrum (and presumably structure) like that of the zirconium analogue $Zr(BH_4)_4$.

Three-centre two-electron hydrogen bridges not involving boron are known in other mixed hydrides. For example, the tetrahydroaluminate $LiAlH_4$ contains tetrahedral AlH_4^- anions bridged by Li^+ cations in tetrahedral holes, the lithium–hydrogen distances being substantially shorter than those in lithium hydride itself (188–200 pm, cf. 204 pm for LiH).[9]

5–4   Some polynuclear hydrogen-bridged transition metal carbonyl hydride complexes[10]

Electron deficient hydrogen bridges also feature in a number of transition metal hydride complexes, such as the binuclear carbonyl hydride anions $[HM_2(CO)_{10}]^-$, where M = Cr, Mo, or W, which are believed to contain linear MHM bridges (see Fig. 5–8a) and may be regarded as derived from the doubly charged anions $[(CO)_5M-M(CO)_5]^{2-}$ by protonation of their single metal–metal bonds. Location of the hydrogen atoms in such complexes is exceedingly difficult—their contributions to the X-ray diffraction pattern are swamped by those due to the heavy metals. How-

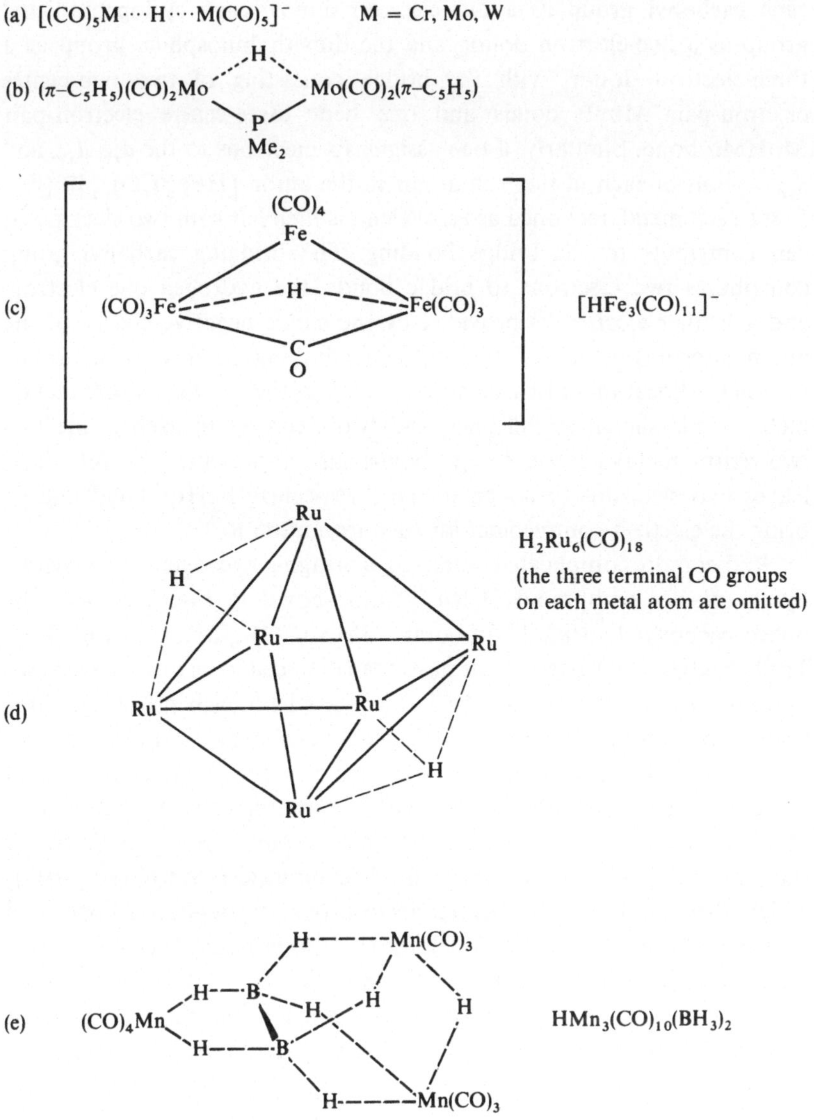

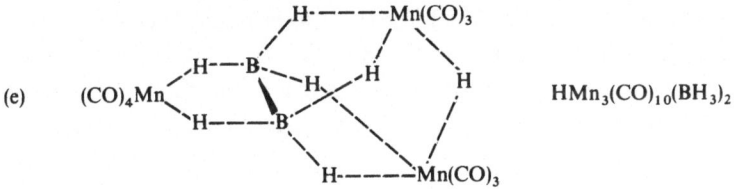

(a) $[(CO)_5M \cdots H \cdots M(CO)_5]^-$ M = Cr, Mo, W

(b) $(\pi-C_5H_5)(CO)_2Mo$ H, P, Me₂ $Mo(CO)_2(\pi-C_5H_5)$

(c) [(CO)₄ Fe, (CO)₃Fe, H, Fe(CO)₃, C O] $[HFe_3(CO)_{11}]^-$

(d) $H_2Ru_6(CO)_{18}$
(the three terminal CO groups
on each metal atom are omitted)

(e) $(CO)_4Mn$ ... $HMn_3(CO)_{10}(BH_3)_2$

Fig. 5–8 Some transition metal carbonyl hydride complexes containing bridging hydrogens

ever, their positions can often be inferred from the arrangement of the other ligands. Compounds in which hydrogen is but one of two or even three bridging atoms between two others are exemplified by the structures shown in Fig. 5–8 b and c. For both of these, as for the anions $[HM_2(CO)_{10}]^-$, an electron count shows one pair of electrons to be available for the MHM bond. Thus in the binuclear molybdenum complex shown in Fig. 5–8b, each molybdenum atom has 18 electrons in its valence shell if one regards each carbonyl group as a two-electron donor, each cyclopentadienyl group as a five-electron donor, and the dimethylphosphino group as a three-electron donor, with the bridge consisting of two two-centre electron-pair Mo–P bonds and one bent three-centre electron-pair MoHMo bond. Similarly, if one assigns six electrons to the d_{xy}, d_{xz}, and d_{yz} orbitals of each of the iron atoms in the anion $[HFe_3(CO)_{11}]^-$ (Fig. 5–8c), each metal, reckoned as zerovalent, is then left with two electrons it can contribute to the bridge bonding. The bridging carbonyl group contributes two electrons to bridge bonds, the hydrogen one electron, and a further electron is provided by the single negative charge of the anion, affording a total of 10 electrons for linking the three metal atoms together. Allocation of two electrons to each of the two two-centre metal–metal bonds shown as full lines, and two electrons to each of the two two-centre metal–carbon bridge bonds also represented by full lines, leaves two electrons to assign to the three-centre FeHFe bond and so bring the electron complement for each iron atom to 18.

Rather more complicated examples of bridging hydrogens are provided by the cluster compound $H_2Ru_6(CO)_{18}$ shown in Fig. 5–8d and the mixed carbonyl hydride hydroborate $HMn_3(CO)_{10}(BH_3)_2$ in Fig. 5–8e. Triply bridging hydrogen atoms have been deduced to be located over two opposite octahedral faces of the former, in which each metal has three terminally attached carbonyl groups which lean away from the two faces in question.[11] For this compound, a description of the bonding in terms of localized three-centre bonds is clearly inappropriate. However, one MnHMn and six MnHB three-centre electron-pair bonds provide an apt rationalization of the structure of the hydroborate $HMn_3(CO)_{10}(BH_3)_2$ which also contains a two-centre electron-pair boron–boron bond, and so uses the total of 16 electrons left for skeletal bonding after six on each metal atom have been assigned to d_{xy}, d_{yz}, and d_{xz} orbitals (i.e. counting $7-6 = 1$ electron as contributed by each metal atom, three electrons as contributed by each boron, and one by each hydrogen to the skeletal bonding).

5–5   Organo-aluminium compounds

The elements whose alkyl or aryl derivatives tend to associate by means of electron deficient MCM bridging groups are shown enclosed by a broken line in Fig. 5–1. Among the Group III elements, aluminium is the only one with a strong tendency to form electron deficient alkyl or aryl bridges. Its tri-alkyls and -aryls R_3Al tend to dimerize (unless the groups R are particularly bulky) by adopting structures $R_2AlR_2AlR_2$ like that of diborane, with two bridging and four terminal groups R. Studies on these dimers have revealed several interesting features of MCM bridges in an environment where they may be compared with terminal M–C links in the same molecule, and indeed with other bridging and terminal M–X links in either the same, or a closely related, molecule.

The bridging unit in dimers $(R_3Al)_2$ consists of a four-membered $(AlC)_2$ ring, as exemplified by the structure of the trimethylaluminium dimer, $(Me_3Al)_2$, shown in Fig. 5–9, which has been studied by X-ray crystallography.[12] This particular structure is one that has been less certainly established than the structures of some other organo-aluminium compounds. The difficulty of locating the hydrogen atoms, and the evident slight distortion from a symmetrical diamond shape of the $(AlC)_2$ ring in this compound allow an alternative interpretation of the X-ray diffraction data in terms of the structure shown in Fig. 5–10, which has a six-membered $(AlCH)_2$ ring.[13] However, this alternative structure has some unsatisfactory features, as outlined below; moreover, there appears to be no infrared or ^1H n.m.r. evidence of hydrogen bridges in trimethylaluminium, and the ^{27}Al nuclear quadrupole resonance (NQR) spectrum, by which the degree of asymmetry in the chemical environment of the aluminium can be found, is better interpreted in terms of the structure shown in Fig. 5–9.[14] (In an NQR experiment, radiation in the radio frequency region is employed to effect transitions among the various orientations of a quadrupolar nucleus in an asymmetric field.)

Other systems for which there is better X-ray crystallographic evidence of AlCAl bridges include the compound $Ph_2NAl_2Me_5$, which contains both bridging carbon and bridging nitrogen atoms[15] (Fig. 5–11) and the compounds $(R_2AlPh)_2$ (R = Ph or Me), both of which contain bridging phenyl groups[16] (Fig. 5–12).

The bonding in an AlC_2Al bridge is intelligible in terms of two closed

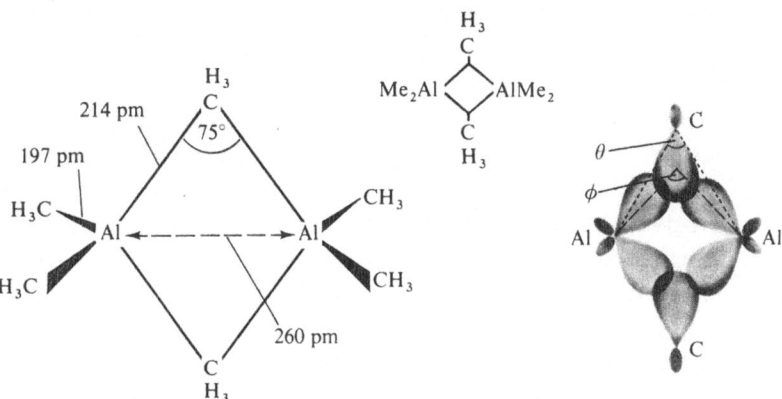

Fig. 5–9 Probable structure of the trimethylaluminium dimer, $(Me_3Al)_2$, and the sp^3 hybrid atomic orbitals (two on each aluminium atom and one on each bridging carbon) that can overlap to form two closed three-centre bonds

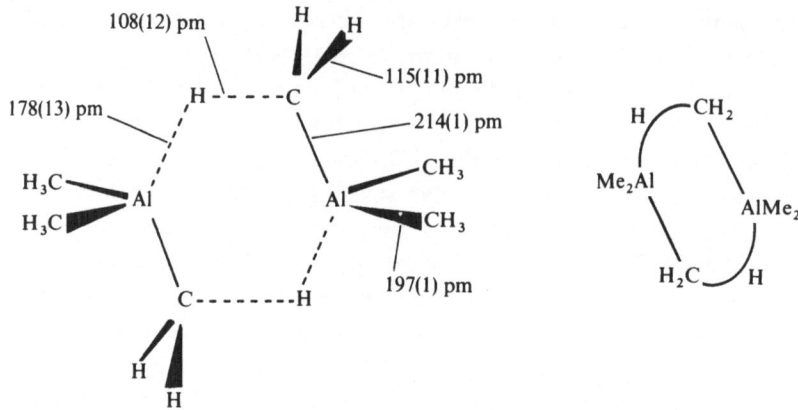

Fig. 5–10   Alternative structure for the trimethylaluminium dimer in terms of which the X-ray crystallographic data have been interpreted

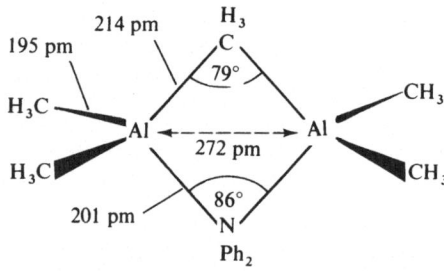

Fig. 5–11   Structure of μ-diphenylamino-μ-methyl-tetramethyl dialuminium, $Me_5Al_2NPh_2$

three-centre bonds resulting from overlap of sp^3 hybrid atomic orbitals (two on each aluminium atom, one on each bridging carbon) as shown in Fig. 5–9 [cf. the closed three-centre BBB bond (Fig. 3–1)]. Good overlap between these orbitals can be expected provided that the bridging AlCAl angle is small— note that this angle (θ in the orbital diagram in Fig. 5–9) must be less than that (ϕ in Fig. 5–9) between the overlapping sp^3 hybrid orbitals of the two metal atoms. The bridging AlCAl angle of 75° is indeed small enough to allow appreciable orbital overlap, and to cause the two metal atoms to be separated by a distance (260 pm) only slightly greater than twice the aluminium single-bond covalent radius (126 pm), which is appropriate in view of the degree of metal–metal bonding implicit in this treatment. By contrast, in dimeric aluminium

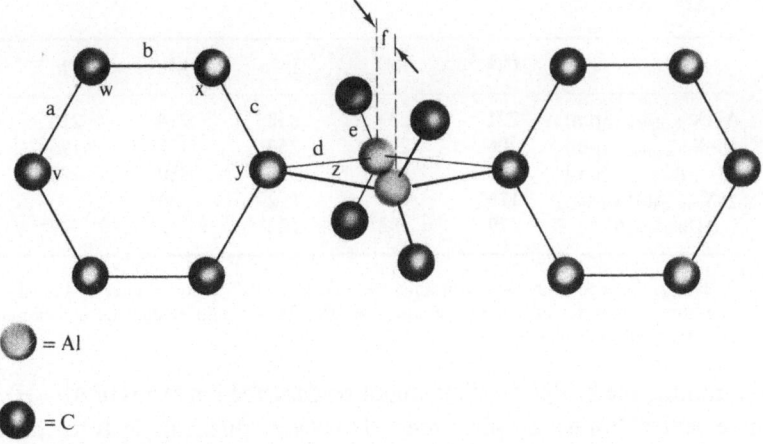

= Al

= C

Interatomic distances (pm) and bond angles:

	a	b	c	d	e	f	v	w.	x	y	z
$Ph_2AlPh_2AlPh_2$	139	140	142	218	196	270	118°	120°	123°	114°	77°
$Me_2AlPh_2AlMe_2$	138	141	142	213	198	268	121°	120°	122°	116°	78°

Fig. 5-12 Skeletal features of the phenyl-bridged aluminium compounds $(Ph_3Al)_2$ and $(Me_2AlPh)_2$

halides, Al_2X_6, in which there are enough electrons to allocate two to each Al–X two-centre bond,

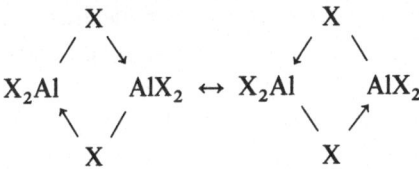

and so no need to invoke metal–metal bonding interactions, the metal–metal distance is significantly greater than that in trimethylaluminium (see Table 5–2), although the chloride and bromide resemble the methyl in having acute bridging AlXAl angles.[17] The compound $Ph_2NAl_2Me_5$ (Fig. 5–11) which has a mixed bridge, has a metal–metal distance appropriate for the AlCAl three-centre bond description, and the bridging angle at nitrogen (86°) significantly exceeds that at carbon (79°). Its bridging Al–N distance is also shorter than the bridging Al–C distance by an amount (*ca* 13 pm) which is roughly twice the difference between the single bond covalent radii of carbon and nitrogen (*ca* 7 pm).[15]

If the structure shown in Fig. 5–10 were correct for dimeric trimethyl-

Table 5–2 Interatomic distances and bond angles in some compounds Al_2X_6 (i.e. $X_2AlX_2AlX_2$)

X	Cl*	Br	I*	Me	Ph
Al–X bridging (pm)	221	233	258	214	218
Al–X terminal (pm)	206	221	253	197	196
Al···Al (pm)	340	321	320	260	270
∠ X_terminalAlX_terminal	118°	115°	112°	124°	—
∠ AlX_bridgingAl	79°	82°	102°	75°	77°

* Data in these columns were obtained from electron-diffraction studies on the vapour; the bond angles are only approximate (±6–$10°$). The remaining data were obtained by X-ray crystallography.

aluminium, the bridge bonding might be described in terms of Al···H···C three-centre bonds as illustrated. However, puzzling features of this structure would be the relatively short metal–metal distance, which is no longer a necessary condition for good overlap of the orbitals involved; the relative length of the ring Al–C bond compared with its terminal analogue, considering that both are formally two-centre electron-pair bonds; and the apparently greater length of the terminal C–H bonds (115 ± 11 pm) than that of the bridging C–H bond (108 ± 12 pm) considering that the *latter* is effectively of order one-half compared with the former.

For all the structures shown with AlCAl bridges (Figs 5–9, 11, and 12), the electron deficient bridging Al–C bonds are, as might be expected, *longer* than the terminal Al–C bonds. That this difference is not entirely due to the electron deficient character of the bridge bonds is apparent from the data in Table 5–2, from which it is seen that bridging Al–X bonds, irrespectively of whether they are regarded as one- or two-electron bonds, are normally longer than terminal Al–X bonds, though the difference is most marked in compounds with electron deficient bridges.

When phenyl groups act as bridging groups, as in the two compounds illustrated in Fig. 5–12, they are inclined approximately at right angles to the plane of the central $(AlC)_2$ ring (actually at 84° to this plane in the case of $(Ph_3Al)_2$), so that the bridging carbon is surrounded by a distorted tetrahedron of two carbon and two aluminium atoms.[16] Indeed, the ring CCC angle at this carbon (angle y in Fig. 5–12) is significantly lower than the hexagonal angle of 120°, and three of the other ring angles likewise deviate from 120°, as if electronic charge were drained from the ring into the Al–C bonds:

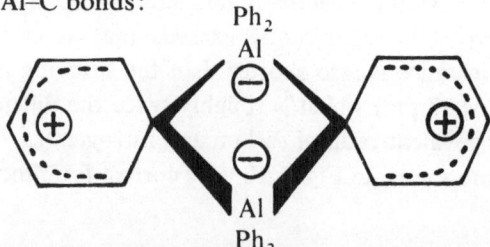

However, as a similar though less pronounced distortion of the terminally-attached phenyl groups is observed in dimeric triphenylaluminium and also in phenyl-gallium and indium compounds, this may be a general feature of phenyl groups attached to electropositive atoms, rather than a peculiarity of electron deficient bridging phenyl.

The structures adopted by the compounds $Ph_2NAl_2Me_5$ (Fig. 5–11) and $(Me_2AlPh)_2$ (Fig. 5–12) show that diphenylamino and phenyl bridging groups are preferred to methyl. Qualitative estimates of the relative bridging capacities of different groups have been made by identification, using ^1H n.m.r. spectroscopy, of the bridging group in mixed alkyls or aryl–alkyls $(R^1R_2^2Al)_2$.[18] Trimethylaluminium itself in solution at $-75°C$ has a spectrum containing two peaks, of relative intensities $2:1$, assignable to the protons of the four terminal and two bridging methyls respectively, although methyl exchange is so rapid at room temperature as to give rise to only one peak. The spectrum of the mixed derivative $(PhC \vdots CAlMe_2)_2$, however, contains a signal due to the aryl protons and only one other peak, irrespective of the temperature, indicating that bridging is through the phenylethynyl groups. The presence of mixed bridges in other systems has been demonstrated similarly—for example, the spectrum of a mixture of $(Me_2AlCl)_2$ and $(Me_2AlBr)_2$ contains three peaks, corresponding to the terminal methyls of the three halogen-bridged species in equilibrium:

$$Me_2AlCl_2AlMe_2 + Me_2AlBr_2AlMe_2 \rightleftharpoons 2Me_2AlBrClAlMe_2$$

Largely as a result of such studies, it has been found that there is a decreasing capacity for bridge formation in the sequence:

$$R_2N > RO > Cl > Br > PhC \vdots C > Ph \sim RCH:CH > Me$$

Bridging by dative two-centre bonds is generally stronger than bridging by three-centre bonds, and follows the sequence expected from the relative Lewis basicities of H_3N, H_2O, HCl, and HBr. Unsaturated groups form three-centre bridges more readily than do saturated groups, which themselves vary in bridging capacity according to the degree of substitution of the bridging carbon. The bridging capacity of alkyl groups as indicated by the degree of association of the tri-alkyls at, or a little above, room temperature decreases in the following sequence:

$$Me > Et > Pr^i > Bu^t$$

This sequence is the reverse of that expected if one considers the relative electron-releasing capacities of hydrogen and methyl groups—the tertiary butyl group, with three electron-releasing methyl groups attached to its central carbon, might have been expected to bridge more strongly than a methyl group, whereas tri-t-butylaluminium, Bu_3^tAl, is actually monomeric. The dominating factor is probably the bulk of the bridging group—the crowding of the dimer increases with the number of methyl substituents on the bridging carbon—though interestingly the observed sequence is that which would be expected if $C \cdots H \cdots Al$ hydrogen bridges were involved, as in Fig. 5–10.

The strong tendency of aluminium to bridge directly through hydrogen is apparent from the structure of aluminium hydride (Fig. 5–2). Organo-aluminium hydrides $(R_2AlH)_n$ bridge through hydrogen in preference to carbon, forming trimers $(n = 3)$, presumably with six-membered $(AlH)_3$ rings, in solution in inert solvents, though the highly viscous nature of the methyl compound $(Me_2AlH)_n$ shows that it probably has a more highly associated structure in the liquid phase. The flexibility of the hydrogen bridges in these compounds, which arises from the spherical symmetry of the hydrogen $1s$ orbitals allowing good overlap at all bridge angles, is reflected in the broadness of their stretching absorptions, near $1800\ cm^{-1}$, in their infrared spectra.

The energy required to cleave the hydride bridges of dialkylaluminium hydrides lies in the range 70–80 kJ/mole per hydride bridge, i.e. very similar to that of the boron analogues. This may be compared with a value of only about 40 kJ/mole per methyl bridge of trimethylaluminium dimer,

$$Me_6Al_2 \rightarrow 2Me_3Al \qquad \Delta H = 83\ kJ/mole$$

and only about 50–60 kJ/mole per halide bridge for the dimeric aluminium halides:

$$Al_2X_6 \rightarrow 2AlX_3 \qquad$$

X:	Cl	Br	I
ΔH:	124	121	102 kJ/mole

At first sight, these values appear to indicate that the strength of the AlXAl bridge decreases in the sequence:

$$X = H > Cl > Br > I > Me$$

and that halide bridges are only marginally stronger than methyl bridges. However, these figures relate to dissociation of dimers, containing tetrahedrally coordinated aluminium, into planar monomers, containing trigonally coordinated aluminium:

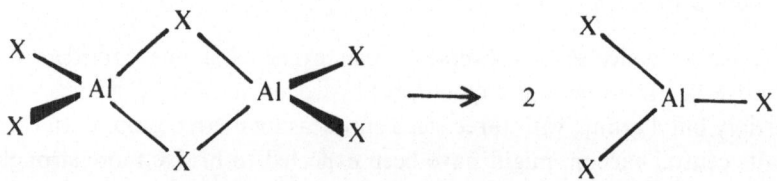

A better guide to the strength of the bridges would be provided by the energy needed to separate the two halves of a dimer Al_2X_6 into *pyramidal* monomers AlX_3:

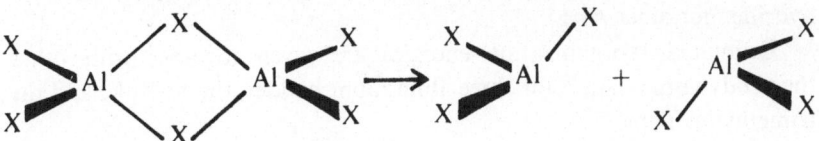

This is because planar monomers differ in energy from hypothetical pyramidal monomers (see Fig. 5–13) by an amount (the 'reorganization

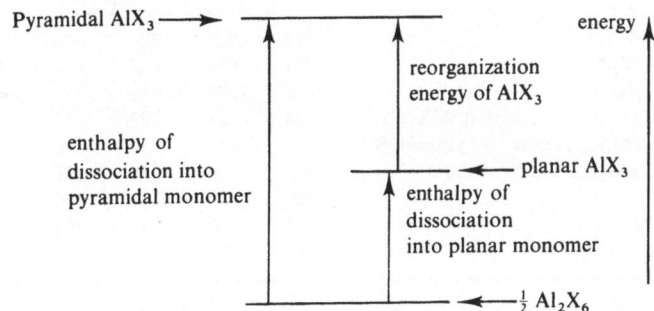

Fig. 5–13 The energy relationship between a planar monomer AlX_3, a hypothetical pyramidal monomer AlX_3, and a dimer Al_2X_6

energy') which may be regarded as composed of the following two major factors:

1. the $(p \leftarrow p)\pi$ Al–X bonding that may be present in the planar monomer but not in the dimer;

2. differences in the Al–X σ-bond energy as the hybridization of aluminium changes from sp^2 to sp^3.

A third factor, the increase in internuclear and interelectronic repulsion forces that accompanies the change from planar to pyramidal geometry, will also contribute to the reorganization energy, but to an extent that may be ignored in a crude treatment.

Neither the reorganization energies themselves, nor these individual factors that contribute to them, can be measured directly, though useful estimates of their values can be made. The first factor, the change in π-bond energy, will be large and significant (of the order of 100 kJ/mole)[19] for the aluminium halides, but zero for such species as AlH_3 or $AlMe_3$ unless hyperconjugation is important in the latter. The second factor, a change in the σ-bond energy with the hybridization of the aluminium, is unlikely to be important if the terminal Al–X bond length does not change going from monomer to dimer, as in the case of trimethylaluminium: the terminal Al–C distance in the dimer Al_2Me_6 is the same as the Al–C distance in the planar monomer $AlMe_3$.[20] The reorganization energy of trimethylaluminium can accordingly be taken as zero to a first approximation. A negligible reorganization energy is similarly expected for the hydride monomer AlH_3.

Estimated reorganization energies are given together with other thermodynamic data[21] for three aluminium halides, the hydride, and the trimethyl in Table 5–3.

Table 5–3 Some thermodynamic data for compounds $(AlX_3)_{1 \text{ or } 2}$

	Cl	Br	I	H†	Me
$-\Delta H_f^\circ(AlX_3)$	583	425	208	—	74
$-\Delta H_f^\circ(Al_2X_6)$	1291	971	517	—	231
$\Delta H_a(Al_2X_6)$	2675	2295	1811	—	8450
$\Delta H_{\text{dissoc.}}(Al_2X_6 \rightarrow 2 \text{ planar } AlX_3)$	124	121	102	150	83
$\Delta H_{\text{reorg.}}(AlX_3; \text{ planar} \rightarrow \text{pyramidal})$	132	117	78	0	0
$\Delta H_{\text{dissoc.}}(Al_2X_6 \rightarrow 2 \text{ pyramidal } AlX_3)$	388	354	258	150	83
$E(Al–X)$	380	325	260	270	260*
$E(AlXAl)$	575	500	385	345	300*
$E(AlXAl)–E(Al–X)$	195	175	125	75	40

Energies throughout are expressed in kJ/mole.

ΔH_f° = enthalpy of formation of the gaseous compound from the elements in their standard states.
ΔH_a = enthalpy of atomization of the gaseous compound, calculated from ΔH_f° by use of the following heats of atomization of the elements:
X = Al, 326·4; Cl, 121·7; Br, 111·9; I, 106·8; C, 716·8; H, 218·0 kJ/g atom.
* Calculated assuming E(C–H) = 415 kJ/mole.
† $\Delta H_f^\circ(AlH_3)$ and $\Delta H_f^\circ(Al_2H_6)$ for gaseous samples are unknown; figures in this column are consistent with the known heat of formation of solid AlH_3 (-46 kJ/mole) and with the known heats of dissociation of dialkylaluminium hydrides $(R_2AlH)_3$ (ca. 75 kJ/mole per H bridge) calculated assuming E(C–H) = 415 kJ/mole.

Adding these to the enthalpies of dissociation of the dimers Al_2X_6 into planar monomers AlX_3 gives values of the enthalpies of dissociation of the dimers into pyramidal monomers, i.e. the energies needed to break two AlXAl bridge units and form two terminal Al–X bonds of the same type as those present in the original dimer. Knowing also the enthalpies of atomization of the dimers Al_2X_6, it is possible to calculate bond energy terms for the two types of bonding unit in the dimers; E(Al–X), the bond energy term for the terminal Al–X bonds, and E(AlXAl), the bond energy term for the three-atom bridging unit Al–X–Al.

The method may be illustrated by considering aluminium chloride as an example. The sum of the bond energy terms for the molecule Al_2Cl_6 is equal to its enthalpy of atomization:

$$4E(Al–Cl) + 2E(AlClAl) = 2675 \text{ kJ/mole}$$

Dissociation of the dimer into pyramidal monomers involves the replacement of two AlClAl bridging units by two terminal Al–Cl bonds

$$2E(AlClAl) - 2E(Al-Cl) = 388 \text{ kJ/mole}$$

Solving these two equations and rounding off the resulting figures, we obtain values of 380 and 575 kJ/mole for E(Al–Cl) and E(AlClAl) respectively.

Similar arguments lead to the values in Table 5–3 of E(Al–X) and E(AlXAl) for the other halides, and (taking E(C–H) to be 415 kJ/mole) to values of E(Al–Me) and E(AlMeAl). The figures in Table 5–3 for Al–H bonds are compatible with the known heat of formation of solid aluminium hydride (which contains three AlHAl bridges per AlH_3 unit) and with the heat of dissociation of AlHAl bridges in organoaluminium hydrides.

In the last row of Table 5–3 are listed values of E(AlXAl)–E(Al–X), the difference between the bond energy terms for the three atom bridging system AlXAl and the terminal two-centre bond Al–X. This energy difference represents the enthalpy of cleavage of one Al–X link in an AlXAl bridge unit. It decreases in the sequence expected from the ^{1}H n.m.r. spectroscopic studies, that is, in the sequence, X = Cl > Br > I > H > Me. The electron pair two-centre dative bridge bonds formed by the halogens are much stronger than the electron deficient bridges formed by hydrogen or carbon. Nevertheless, the two Al–X two-centre links in an AlXAl bridge in which X = halogen are significantly *weaker* than terminal Al–X bonds; the difference is most marked for the chloride, for which the ratio E(AlClAl): E(Al–Cl) is only about 3:2, not 2:1 as would be expected if both bridging and terminal Al–Cl links were of comparable strength. These bond energy terms thus reinforce the conclusion drawn from an inspection of the relative lengths of bridging and terminal aluminium–halogen bonds given in Table 5–2.

5–6   Some reactions of aluminium alkyls

Some typical reactions of aluminium alkyls are listed in Table 5–4. Many of these closely resemble those of diborane, and have been classified accordingly (cf. Table 4–4). Major features again include the susceptibility to attack by Lewis bases, protic acids, unsaturated functional groups, and oxygen. The alkyls are useful alkylating and reducing agents, and are used in many catalyst systems, particularly for olefin oligomerization or polymerization. There is space here to note only a few features. More detailed treatments are to be found in refs 22–24.

The trialkyls of aluminium are highly reactive species. All three of their Al–C bonds are cleaved by water or alcohols in reactions which at room temperature may be explosively violent, though which when carried

Table 5–4   Some reactions of aluminium trialkyls $(R_3Al)_{1\ or\ 2}$

1 Formation of adducts with donor molecules

$$R_6Al_2 \underset{-L}{\overset{+L}{\rightleftharpoons}} R_2(L)Al \cdots R \cdots AlR_3 \underset{-L}{\overset{+L}{\rightleftharpoons}} 2L,AlR_3$$

$R_6Al_2 + 2L \rightarrow 2L,AlR_3$ (L = Me$_3$N, Me$_3$P, Me$_2$O, Me$_2$S, Me$_2$Se, Me$_2$Te in order of decreasing donor strength; also many other Lewis bases, R$_2$CO, RCN, etc.)

$R_6Al_2 + X^- \rightarrow [R_3AlXAlR_3]^-$ (X = H, halogen, OR, NR$_2$)

$R_6Al_2 + 2X^- \rightarrow 2[AlR_3X]^-$ (X = H, CN, halogen, OR, R)

2 Reactions with compounds $H(\delta +)$—$X(\delta -)$ to eliminate hydrocarbon RH

$$R_6Al_2 + HX \rightarrow RH + R_2Al \overset{\displaystyle R}{\underset{X}{\diamond}} AlR_2 \qquad \text{(X = NR}_2\text{, OR, halogen in order of increasing ease of elimination of RH)}$$

$$R_6Al_2 + 2HX \rightarrow 2RH + R_2Al \overset{X}{\underset{X}{\diamond}} AlR_2 \quad \text{or} \quad R_2Al \overset{X \rightarrow AlR_2}{\underset{X-AlR_2}{\diamond}} X$$

(X = NR$_2$, PR$_2$, AsR$_2$, OR, SR, halogen; dimeric products (R$_2$AlX)$_2$ are commoner than trimers (R$_2$AlX)$_3$, which result only if R and X are relatively small)

$R_6Al_2 + $ excess $HX \rightarrow 6RH + 2AlX_3$ (X = OH, OR, Cl, Br; the first two are useful in quantitative analysis)

3 Reactions with unsaturated functional groups

$$\diagup Al\!-\!R \xrightarrow[ca.\ 100°;\ \text{high pressure}]{CH_2:CH_2} \diagup Al(C_2H_4)_nR \ (R = Et, Pr, etc., but not Me)$$

(as monomer, R$_3$Al)

$$Et_6Al_2 + \text{a transition metal halide} \xrightarrow[\substack{\text{heptane suspension} \\ \text{room temperature and} \\ \text{pressure}}]{CH_2:CHR} \text{stereoregular polyolefin}$$
(for example, TiCl$_3$)
(Ziegler–Natta catalysts)

$$\diagup Al\!-\!R^1 + R^2C\colon N \!-\!\!\!\begin{cases} \xrightarrow{\text{heat}} \diagup AlN\colon CR^1R^2 \ \ \text{(for example, } R^1 = \text{Me, Ph;} \\ \hphantom{\xrightarrow{\text{heat}} \diagup AlN\colon CR^1R^2 \ \ } R^2 = \text{Ph, Bu}^t \text{)} \\[12pt] \xrightarrow{\text{-olefin}} \diagup AlN\colon CHR^2 \quad (R^1 = \text{Et or higher alkyl)} \end{cases}$$

(Similar additions, with or without loss of olefin, occur to many other unsaturated compounds, for example, R$_2$CO, RNCO, RNCS, CO$_2$)

4 Exchange reactions

$MX_n + $ excess $R_6Al_2 \rightarrow R_nM + AlX_3$ (M = B, Ga, Si, Ge, Sn, etc. X = halogen, OR)

$R_6^1Al_2 + R_6^2Al_2 \rightarrow R_5^1R^2Al_2, R_4^1R_2^2Al_2,$ etc.

$R_6Al_2 + Al_2X_6 \rightarrow R_4Al_2X_2, R_3Al_2X_3, R_2Al_2X_4,$ etc. (X = Cl, Br)

$\diagup Al\!-\!R^1 + R^2COX \rightarrow \diagup AlX + R^1R^2CO$ (X = halogen)

$\diagup Al\!-\!CH_2CH_2R^1 + R^2X \rightarrow \diagup AlX + R^1CH_2CH_2R^2$ (or $R^1CH\colon CH_2 + R^2H$)

5 Oxidative cleavage of Al–R bonds

$R_6Al_2 + excess\ O_2 \rightarrow Al_2O_3 + xH_2O + yCO_2$

(the lower alkyls may spontaneously inflame in air)

6 Reduction of aluminium alkyls

$2R_6Al_2 + 3Na \rightarrow 3NaAlR_4 + Al$

7 Thermal decomposition of aluminium alkyls

$$\diagup AlCH_2CHR^1R^2 \underset{\text{moderate heat}}{\overset{}{\rightleftharpoons}} \quad Al–H + CH_2:CR^1R^2$$

(this reaction is relevant to the reducing properties of aluminium alkyls, to their reactions with olefins and other unsaturated substances, and (in reverse) to the synthesis of aluminium alkyls from aluminium hydrides)

$R_6Al_2 \xrightarrow{ca.\ 300^\circ} Al + H_2 + \text{hydrocarbons}$

(a reaction suitable for coating surfaces with aluminium)

out at low temperature in solution are valuable in quantitative analysis if the hydrocarbon liberated is collected and measured:

$$1/n\,(R_3^1Al)_n + 3R^2OH \rightarrow Al(OR^2)_3 + 3R^1H$$

The alkyls should be handled either under a dry inert atmosphere or under vacuum, as they are also sensitive to oxygen—the lighter alkyls may inflame spontaneously in air. Reactivity normally decreases with the number of alkyl groups per aluminium, i.e. in the sequence $R_3Al >$ $R_2AlX > RAlX_2$, where X is an electronegative atom or group.

Alkyls with hydrogen attached to the β-carbon atom tend to lose olefin reversibly when heated, generating Al–H bonds:

$$\diagup AlCH_2CH_2R \rightleftharpoons \diagup AlH + CH_2:CHR$$

Since Al–H bonds undergo the same types of reaction as do Al–C bonds, only even more readily, this means that reactions requiring heat, or exothermic reactions in which spontaneous heating occurs, may lead to products derived from both Al–C and Al–H starting materials. For example, unsaturated compounds may give products of insertion into both Al–C and Al–H bonds:

$$\diagup AlCH_2CH_3 \qquad \qquad \diagup AlOCR_2Et$$
$$\updownarrow \qquad \xrightarrow{R_2CO} \qquad +$$
$$\diagup AlH + CH_2:CH_2 \qquad \diagup AlOCHR_2 + CH_2:CH_2$$

A further complication arises if the unsaturated compound contains hydrogen attached to an α-carbon atom, when protic acid reaction is possible, as in the enolic reaction of methyl ketones:

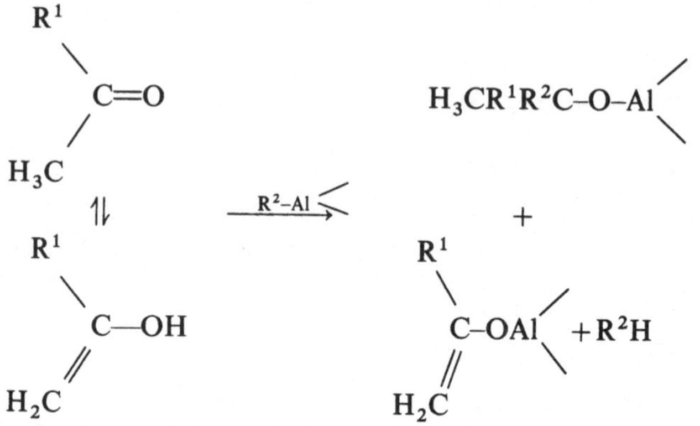

$$\begin{array}{c}
R^1 \\
\diagdown \\
\qquad C{=}O \\
\diagup \\
H_3C
\end{array}
\qquad \xrightarrow[\;R^2\text{-Al}\diagdown\;]{}\qquad
H_3CR^1R^2C\text{-O-Al}\diagup\diagdown \\[2mm]
\Big\updownarrow \qquad\qquad\qquad\qquad\qquad +$$

$$\begin{array}{c}
R^1 \\
\diagdown \\
\qquad C{-}OH \\
\diagdown\!\!\diagdown \\
H_2C
\end{array}
\qquad\qquad\qquad
\begin{array}{c}
R^1 \\
\diagdown \\
\qquad C\text{-OAl}\diagup\diagdown \;\; +R^2H \\
\diagdown\!\!\diagdown \\
H_2C
\end{array}$$

Coordination complexes L,Al_2R_6 or L,AlR_3, in which an AlRAl bridge has probably been cleaved for each base molecule L attached to aluminium, feature as intermediates in many reactions. For example, the reaction of trimethylaluminium with benzophenone in benzene, which after hydrolysis affords 1-diphenylethanol, Ph_2CMeOH, occurs via the adduct $Ph_2CO,AlMe_3$ which is believed to react subsequently by one of two possible paths, depending on whether an excess of trimethylaluminium is present (see Fig. 5–14).[25] In these and many other reactions of aluminium

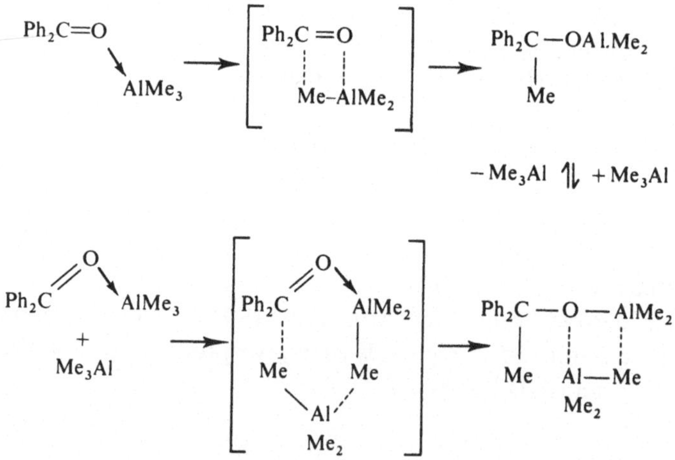

Fig. 5–14   Probable intermediates in trimethylaluminium–benzophenone reactions

alkyls, the alkyl groups or hydrogen atoms transferred probably play a bridging role in the intermediates involved.[26]

Unlike the boron atoms of diborane, the aluminium atoms of the dimeric trialkyls $(R_3Al)_2$ can expand their coordination numbers above four, and apparently do so in the intermediates of some exchange reactions. For example ^1H n.m.r. studies of the exchange of methyl groups between trimethylaluminium (as monomer) and dimethyl(phenylethynyl)aluminium have indicated that five-coordinate species may be involved:[27]

The involvement of five-coordinate intermediates would also account for the readier exchange of substituents between adducts L,AlR_3^1 and L,AlR_3^2 than occurs with their boron counterparts, and also for the greater ease of hydrolysis of such adducts than their borane analogues:

$$L,AlR_3 + H_2O \rightarrow [H_2O,AlR_3,L] \rightarrow RH + HOAlR_2,L \rightarrow \text{etc.}$$

Five-coordinate adducts of the type $L^1L^2AlR_3$, represented by the intermediate above are not normally isolable except in the special case where R = H. The bis(trimethylamine) adduct $(Me_3N)_2AlH_3$ for example has a trigonal bipyramidal coordination of three equatorial hydrogens and two axial nitrogens about the metal (by X-ray crystallography).

The industrially important 'growth' reactions of aluminium alkyls with olefins, whereby oligomers or polymers of the olefin may be generated by successive insertions into the Al–C bonds,

$$\diagdown Al–R \xrightarrow{CH_2:CHR} \diagdown AlCH_2CHR_2 \xrightarrow{CH_2:CHR} \diagdown Al(CH_2CHR)_2R, \text{etc.}$$

involve *monomeric* species R_3Al. A reaction involving dimers R_6Al_2 along the lines described as plausible for the hydroboration of olefins by the di-isopinocampheylborane dimer, $(isopinocampheyl_2BH)_2$ (see Fig. 4–12) would be extremely unlikely, since the approach of the olefin would be blocked by the bridging alkyl group. The reaction may well occur via an olefin–aluminium alkyl complex.

$$R_3^1Al + R^2CH:CH_2 \rightarrow \begin{bmatrix} \overset{\delta+}{R^2}CH = \overset{\delta-}{CH_2} \\ | \\ \underset{\delta-}{R^1} - \underset{\delta+}{AlR_2^1} \end{bmatrix} \rightarrow \begin{array}{c} R^1R^2CHCH_2 \\ | \\ AlR_2^1 \end{array}$$

Evidence for weak olefin → aluminium interactions has come from a study of 4-alkenylaluminium compounds $Bu_2^iAl(CH_2)_3CH:CHR$, in which the alkenyl side chain is long enough to allow intramolecular metal–olefin interactions, the presence of which is indicated by a low C=C stretching frequency compared with that of the parent alkene $RCH:CH(CH_2)_2CH_3$.[28] The olefin stretching frequency increases on addition of ether which destroys the metal–olefin interaction, for example,

$$\begin{array}{c} Bu_2^iAl\!\!-\!\!-\!\!-\!\!-\!\!CH_2 \\ \uparrow \qquad \qquad \diagdown \\ \qquad \qquad \qquad CH_2 \xrightarrow{Et_2O} Et_2O,Al(Bu_2^i)(CH_2)_3CH:CHMe \\ MeCH\!\!=\!\!CH\!\!-\!\!CH_2 \diagup \end{array}$$

$\nu(C:C)$, 1635 cm^{-1} $\nu(C:C)$, 1658 cm^{-1}

When aluminium alkyl–transition metal halide mixtures are used for the stereoregular polymerization of olefins (Ziegler–Natta catalysis), the function of the aluminium alkyl is believed to be to furnish transition metal alkyl bonds into which olefin molecules can successively insert, for example,

$$\geqslant TiCl \xrightarrow[(-ClAl\leqslant)]{+EtAl\leqslant} \geqslant TiEt \xrightarrow{RCH:CH_2} \geqslant TiCH_2CHREt \xrightarrow{RCH:CH_2}$$

$$\geqslant Ti(CH_2CHR)_2Et, \text{ etc.}$$

A further difference between the aluminium alkyls and diborane is found in the response of the alkyls to conditions which convert diborane into higher boranes, since there are no aluminium analogues of the higher boranes. For example, the metal, hydrogen, and assorted hydrocarbons result from thermal decomposition of the alkyls above about 300°, and the metal is also one of the products in the reduction of aluminium alkyls by sodium:

$$2R_6Al_2 + 3Na \rightarrow 3NaAlR_4 + Al$$

Apart from these features, however, there remains a close correspondence between the reactions of aluminium alkyls and those of diborane, as is apparent from the various examples in Table 5–4. The low energy of Al–C bonds, their polarity $Al(\delta+)$—$C(\delta-)$, and the Lewis acidity of the trialkyls AlR_3 provide the driving force and mechanism for their replacement by stronger Al–X bonds.

5–7 Alkyl or aryl bridges involving gallium, indium, and thallium

The heavier elements of Group III have relatively little tendency to form electron deficient bridges involving carbon atoms. Their trimethyl derivatives Me_3M, for example, like trimethylborane, are monomeric in the gas phase and probably also in the liquid. The uniquely dimeric state of trimethylaluminium is reflected by its anomalously high boiling point:

Trimethyl derivative: Me_3B $(Me_3Al)_2$ Me_3Ga Me_3In Me_3Tl
b.p. (°C):   −22   126   56   136   147
                                                                                                    (extra-
                                                                                                    polated)

Nevertheless, X-ray crystallographic studies on the indium[29] and thallium[30] compounds have shown the trigonal planar molecules to be so aligned as to associate feebly, giving a distorted trigonal bipyramidal coordination about the metal (see Fig. 5–15). Weak interactions between

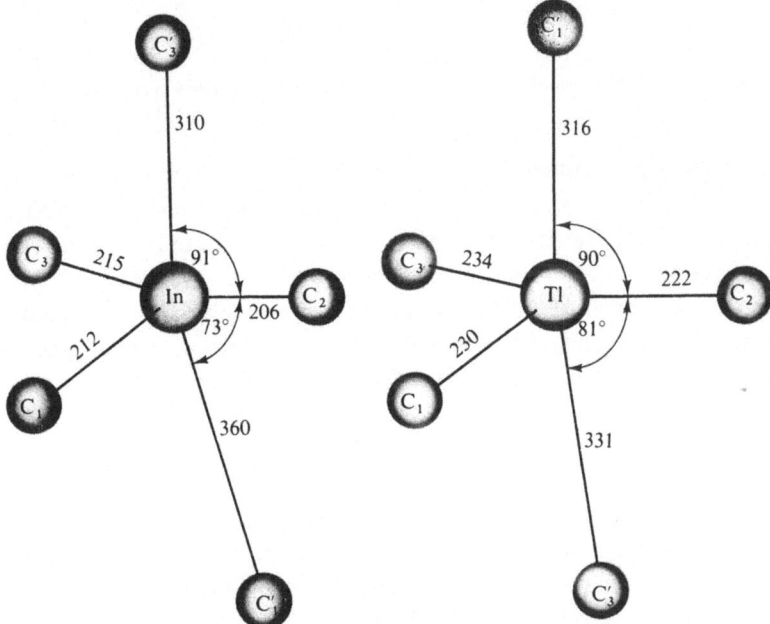

Interatomic distances in pm

Fig. 5–15   The arrangement of carbon atoms about the metal in crystalline Me_3In and Me_3Tl

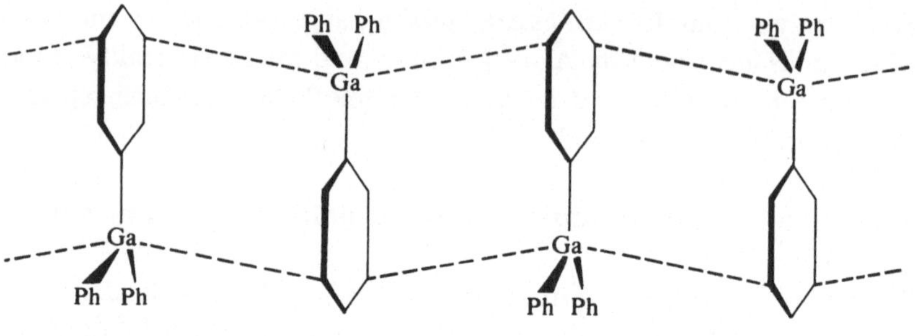

Ga–C interatomic distances (pm):
Intramolecular; 195, 197   Intermolecular; 342

Fig. 5-16 Weak intermolecular gallium-*meta*-carbon interactions in crystalline triphenylgallium

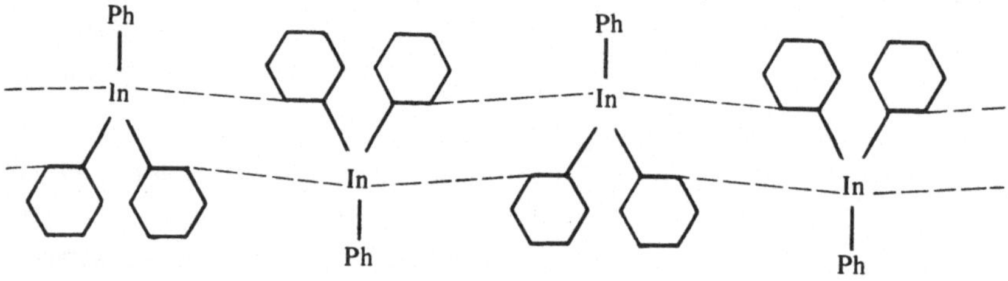

In–C interatomic distances (pm):
Intramolecular; 211, 215   Intermolecular, 307

Fig. 5-17 Weak intermolecular indium-*ortho*-carbon interactions in crystalline triphenylindium

molecules of two triphenyl derivatives are shown in Figs 5–16 and 5–17. In contrast to the type of bridging that occurs in triphenylaluminium, triphenylgallium and triphenylindium associate by interactions between the metal and an *ortho* or *meta* carbon of an adjacent molecule.[31] It is possible that similar feeble bonding interactions remain to be discovered in arylboron compounds. Electron deficient boron–carbon–boron links are otherwise found in carboranes (see Chapter 6).

5-8 Organo-beryllium and -magnesium compounds

The strong tendency for the metal to become four-coordinate noted for organo-aluminium compounds is also a feature of the organometallic chemistry of beryllium and magnesium. Four-coordination of the metal (and use of its four valence shell *s* and *p* orbitals) was achieved in the case of organo-aluminium compounds R_3Al simply by dimerization, and involvement of one-third of their organic groups in AlCAl bridges. Use of the corresponding four valence shell orbitals of beryllium and magnesium in their organo derivatives R_2M requires *all* their organic groups to be involved in MCM bridges. This is indeed the case for the dimethyl derivatives, which have polymeric structures containing MC_2M bridges, the shape of which closely resembles that of their aluminium counterpart with a particularly acute angle at the bridging carbon (Fig. 5–18).[32,33] The resemblance is underlined by the structure adopted by the mixed metal derivative $Mg(AlMe_4)_2$ (Fig. 5–19), in which are seen the acute bridge angle at carbon, the short metal–metal distance, and the greater length of bridging than of terminal Al–C distances that characterize the structure of trimethylaluminium.[34]

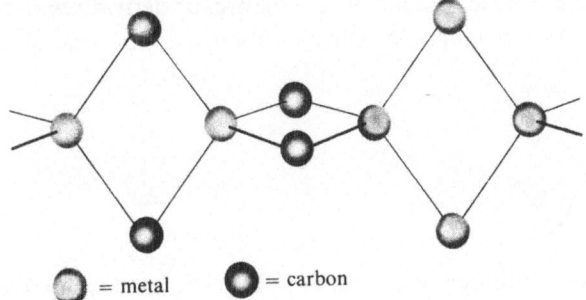

⬤ = metal   ⬤ = carbon

Interatomic distances and angles:

$(Me_2Be)_n$:   Be–C, 193 pm; Be–Be, 209 pm; $\angle$ BeCBe, 66°

$(Me_2Mg)_n$:   Mg–C, 224 pm; Mg–Mg, 272 pm; $\angle$ MgCMg, 75°

Fig. 5–18 Skeletal structure of the dimethylberyllium and dimethylmagnesium polymers

The polymeric nature of the simple beryllium or magnesium dialkyls or diaryls makes them less convenient substances for study than aluminium trialkyls—they have low volatility and low solubility in inert solvents. However, the relative bridging capacities found in aluminium chemistry hold also in Group II. The bulkier dialkyls are oligomers or even monomers. Di-tertiary-butylberyllium, Bu_2^tBe, for example is volatile, monomeric, with a Be–C bond length of 170 pm in the linear CBeC skeleton.

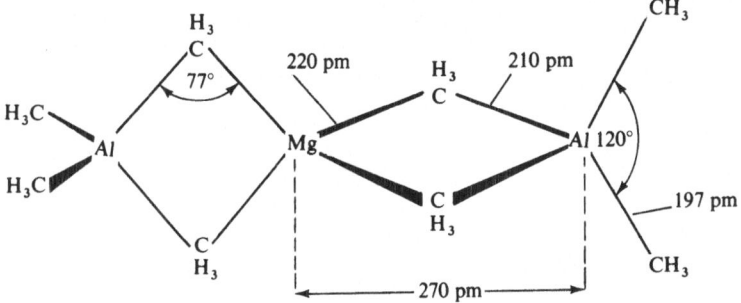

Fig. 5–19 The structure of $Mg(AlMe_4)_2$

Unsaturated organic groups form stronger bridges than do saturated groups, though association through three-centre hydrogen bridges or through dative two-centre bonds from such elements as nitrogen, oxygen, or halogens is preferred.

5–9　Alkali metal alkyls

The alkali metal alkyls RM, having only one alkyl group per metal atom would have a coordination number of only two for the metal (and require only two metal atomic orbitals for bonding) if they associated through MC_2M bridges of the type common to the organo derivatives of aluminium, beryllium, and magnesium already discussed, for example, forming dimers:

$$\ce{>C-<_{M}^{M}>-C<}$$

Dimeric organo-lithium compounds $(RLi)_2$ are, however, extremely rare, except as reactive intermediates in reactions of higher oligomers; stable examples include certain aryl derivatives in ethereal solution. Instead, and evidently in order to make better use of the orbitals available, most organo-lithium compounds adopt tetrameric or even hexameric structures $(RLi)_4$ or $(RLi)_6$ in which the bridging carbon atoms characteristically bridge three, rather than two, metal atoms.[35] Typically, the bridging

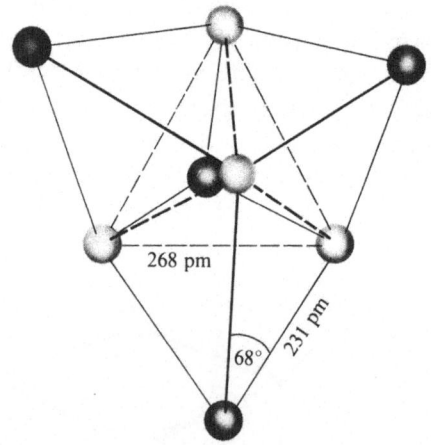

268 pm

68°　231 pm

◐ = Li　　● = C

Fig. 5–20　The Li_4C_4 skeleton of the methyl-lithium tetramer viewed approximately along one three-fold axis

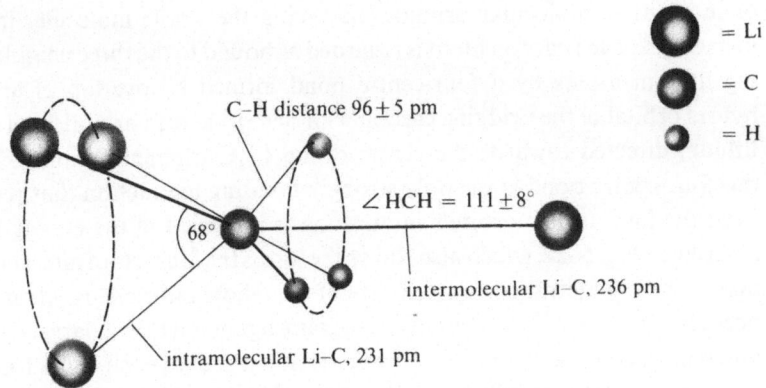

C–H distance 96 ± 5 pm

$\angle$ HCH = $111 \pm 8°$

68°

intermolecular Li–C, 236 pm

intramolecular Li–C, 231 pm

= Li
= C
= H

(a) The effectively seven coordinate environment of each carbon atom

intramolecular Li–Li, 268 pm

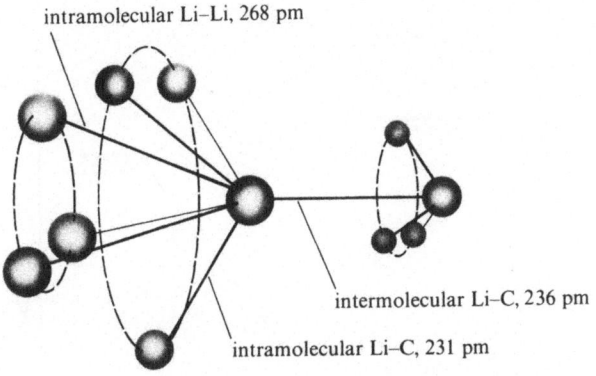

intermolecular Li–C, 236 pm

intramolecular Li–C, 231 pm

(b) The effectively seven coordinate environment of each lithium atom

Fig. 5–21 Details of the environments of the carbon and lithium atoms of crystalline methyl-lithium

carbon atom is located over an equilateral triangle of three lithium atoms, as in the methyl–lithium tetramer, $(MeLi)_4$, the structure of which (see Fig. 5–20) has been deduced from X-ray powder data.[36] In this tetramer, the four lithium atoms are at the corners of a tetrahedron of edge 268 pm, shown by broken lines in Fig. 5–20. The carbon atoms are located over the centres of the four tetrahedral faces, so that they in turn are at the corners of a second tetrahedron which, together with the first, gives rise to a distorted cubic molecular skeleton (shown by full lines in Fig. 5–20). In the lattice, these distorted cubes are so packed as to bring each metal atom of one very close to (within about 236 pm of) one carbon atom of a neighbouring tetramer (see Fig. 5–21).

Within the tetramer, the bonding may be treated either in terms of localized, four-centre (Li_3C) bonds over each face of the Li_4 tetrahedron, or in terms of molecular orbitals embracing the whole molecule. In the former case, each carbon atom is regarded as bound to the three neighbouring lithium atoms by a four-centre bond formed by overlap of an sp^3 hybrid orbital of the bridging carbon with one sp^2 hybrid orbital from each lithium directed towards the centre of the Li_3C trigonal pyramid. That this four-centre bond is the only strongly bonding interaction that results from the four atomic orbitals in question is indicated in the energy level diagram in Fig. 5–22, which also shows the forms the four sets of interactions take. When the tetrahedral symmetry of the whole molecule is taken into account, the orbital energy level diagram again reveals a large energy difference (Weiss and Lucken[36] estimate some 5 eV, i.e. *ca* 480 kJ/mole) between the highest filled and lowest unfilled molecular orbitals (see Fig. 5–23). In this diagram the four orbitals of lowest energy, one of A_1 and three of T_2 symmetry, that result from interactions between the atomic

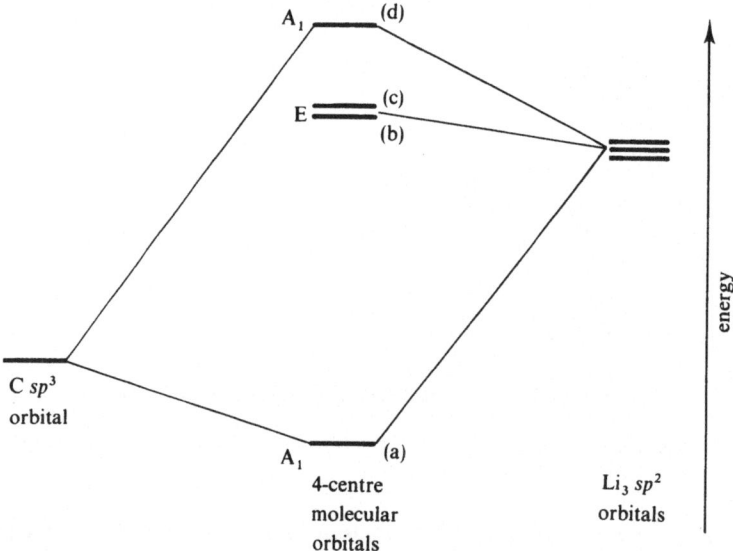

Fig. 5–22 The relative energies and symmetries of the localized 4-centre molecular orbitals that result from interactions between the carbon sp^3 and lithium sp^2 hybrid atomic orbitals pointing towards the centre of an Li_3C trigonal pyramid (of local C_{3v} symmetry)

orbitals of four lithium atoms at the corners of a regular tetrahedron, are stabilized by interaction with the sp^3 hybrid orbitals of the bridging carbons. It is in these orbitals that the eight bonding electrons are accommodated. Their strongly metal–metal bonding character accounts for the relatively short metal–metal distance in methyl-lithium (268 pm; cf. 267 pm in the diatomic molecule Li_2, and 304 pm in lithium metal), also for the persistence of the Li_4 cluster in the fragmentation of methyl-lithium in the mass-spectrometer (methyl groups are lost far more readily than lithium atoms), and for the persistence of the tetrameric state of association even in donor solvents. In principle, one donor molecule can be accommodated on each lithium atom *without* drastically weakening the cluster bonding. Four of the orbitals shown as antibonding in Fig. 5–23 (one of A_1 and three of T_2 symmetry) would be lowered in energy, and become bonding orbitals, through interaction with the orbitals of ligands

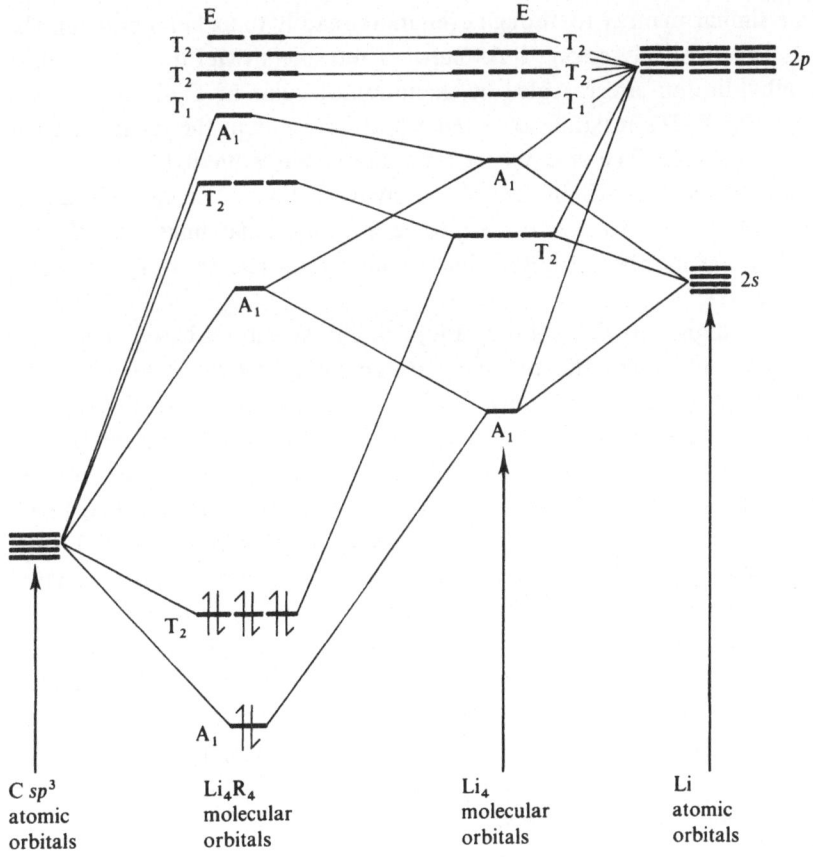

Fig. 5–23 Orbital energy level diagram for $(RLi)_4$

along the three-fold axes. That there will be a total of eight bonding orbitals if a ligand atom is brought along the three-fold axis to within bonding distance of each lithium atom can be seen alternatively by use of a localized bond approach, considering each lithium atom as sp^3 hybridized, with one sp^3 hybrid orbital pointing outwards from the Li_4 tetrahedron along the three-fold axis. If this orbital is used for a coordinate link to a Lewis base L along this axis, there remain on each lithium the three sp^3 hybrid orbitals pointing towards the centres of the Li_4 tetrahedral faces from which to construct the cluster orbitals previously treated as resulting from the use of sp^2 hybrid orbitals. The tetrahedral complex, $(RLiL)_4$, that results is isoelectronic with tetraboron tetrachloride, B_4Cl_4. This ability of the alkyl-lithium tetramer to accommodate four donor molecules without cleavage may be contrasted with the behaviour of the trimethyl-aluminium dimer, which is cleaved by donor molecules L in forming adducts $L,AlMe_3$. In other respects (sensitivity to oxygen, moisture, protic acids, and unsaturated groups, for example), lithium alkyls react in a similar manner to, though even more readily than, aluminium alkyls.

The relatively strong links *between* tetramers $(MeLi)_4$ in crystalline methyl-lithium are revealed by the in*ter*molecular Li–C distance of 236 pm (Fig. 5–21), insignificantly (*ca* 5 pm) longer than the in*tra*molecular Li–C distance. The strength of these links causes methyl-lithium to be involatile and sparingly soluble in hydrocarbons. Directed along the three-fold axes of the tetramers, these intermolecular links raise the co-ordination number of both lithium and carbon effectively to *seven* (see Fig. 5–21).

The higher alkyls of lithium adopt similar structures based on clusters of metal atoms with the invariably bridging alkyl groups located over the triangular faces of these clusters. Tetramers like that of methyl-lithium are particularly common, though the bulkier the alkyl group the less the interaction between tetramers in the crystal, so that *t*-butyl-lithium, $(Bu^tLi)_4$, is readily soluble, as the tetramer, in hydrocarbons, and is much more volatile than methyl-lithium, being sublimable at $70°/0.1$ mm. Ethyl-lithium, which in the crystal consists of tetramer units associated into strips (rather than three-dimensionally like methyl-lithium), dissolves in hydrocarbons as a hexamer $(EtLi)_6$, a species also present in gaseous ethyl-lithium. A possible structure for the hexamer is shown in Fig. 5–24.[35] This consists of a distorted octahedral cluster of lithium atoms, with the bridging alkyl groups over six of the triangular faces, the un-bridged faces being opposite each other.

Among the heavier alkali metals, sodium forms organo derivatives intermediate in character between covalent (electron deficient) and ionic. Thus whereas methyl-sodium crystallizes in tetramers, like methyl-lithium, ethyl-sodium adopts a double layer structure in which the

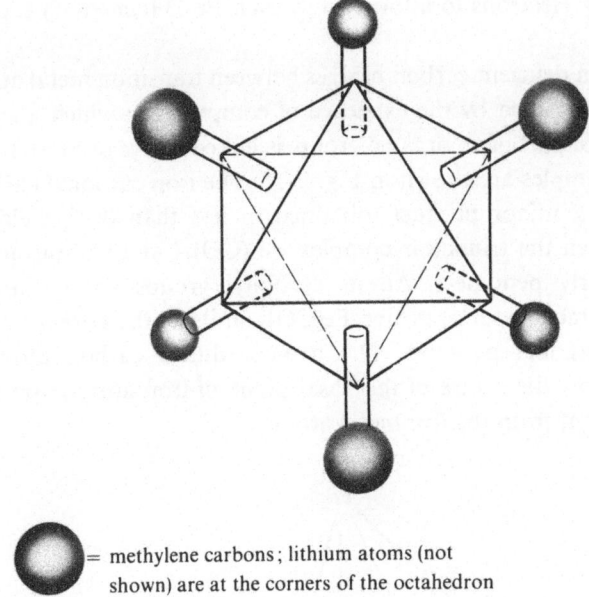

= methylene carbons; lithium atoms (not
  shown) are at the corners of the octahedron

Fig. 5–24   Structure suggested for the ethyl–lithium hexamer, showing the methylene carbon atoms over six of the faces of a distorted Li_6 octahedron

methylene groups of the alkyl residues have a trigonal-pyramidal array of sodium atoms/ions as near-neighbours.[37] Methylpotassium adopts a nickel arsenide type structure, in which methyl groups, effectively as the anions CH_3^- and probably with the expected pyramidal geometry (cf. the isoelectronic ammonia molecule), are surrounded by a trigonal prismatic array of potassium ions.[38]

5–10   Some carbide structures

It is possible that some transition metals may form electron deficient alkyl bridges like those formed by the main group metals discussed above, though this has yet to be demonstrated. Their uncoordinated alkyls MR_n are relatively thermally unstable substances, and few can be isolated and studied at room temperature. It was at one time thought (on the basis of an X-ray crystallographic study) that tetramethylplatinum existed as the tetramer $(Me_4Pt)_4$ with an $(MePt)_4$ skeleton like that of methyl-lithium (Fig. 5–20) with a distorted cubic arrangement of four platinum atoms and four triply bridging methyl groups, but with three terminal methyl groups also attached to each metal atom. However, it later became apparent that the crystals studied were not tetramethylplatinum, but the hydroxide $(Me_3PtOH)_4$, with triply bridging *oxygen* atoms. This compound contains enough bonding electrons to allow two to each Pt–O (*not* Pt–C) edge of the cube.

That electron deficient carbon bridges between transition metal atoms can occur is underlined by the existence of complexes in which a single carbon atom, not part of an organic group, is linked to five or more metal atoms. Two examples are shown in Fig. 5–25. The iron carbonyl carbide $Fe_5(CO)_{15}C$ is a minor product (obtained in less than 0·5% yield) of reactions between the trinuclear complex $Fe_3(CO)_{12}$ and various acetylenes, particularly pentyne-1. All its carbonyl groups are terminally coordinated to metal atoms, in five $Fe(CO)_3$ units at the corners of an equilateral tetragonal pyramid. The five-coordinate carbon atom is located just below the centre of the basal plane of iron atoms, approximately equidistant from the five metal atoms.[39]

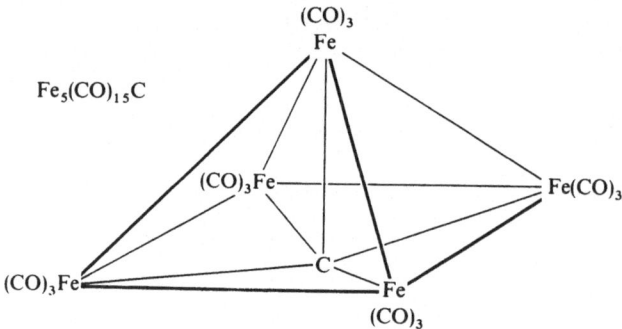

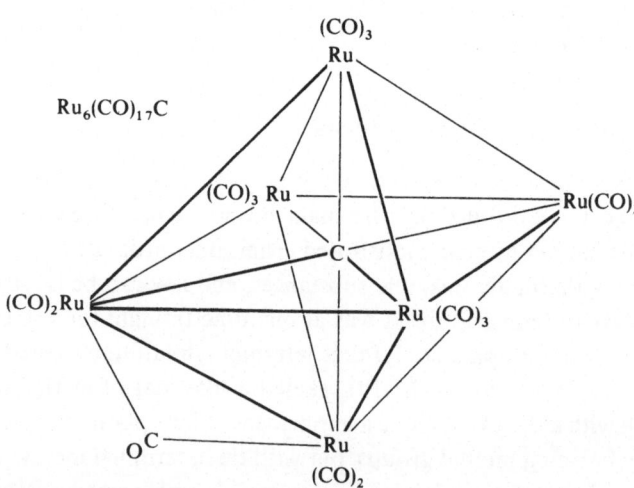

Fig. 5–25 Structures of the carbonyl carbides $Fe_5(CO)_{15}C$ and $Ru_6(CO)_{17}C$

The other structure illustrated in Fig. 5–25 is that of the ruthenium carbonyl carbide $Ru_6(CO)_{17}C$, obtained by thermal decomposition of the trinuclear complex $[Ru(CO)_4]_3$.[40] Its carbide carbon atom is at the centre of an octahedral metal cluster. Sixteen of its carbonyl groups are terminally attached to ruthenium atoms (there are four $Ru(CO)_3$ and two $Ru(CO)_2$ units). The seventeenth carbonyl group acts as a bridging group between the two $Ru(CO)_2$ units. This complex is isoelectronic with the carbonyl hydride $H_2Ru_6(CO)_{18}$ shown in Fig. 5–8d, and its bonding is likewise better treated using a qualitative molecular orbital approach rather than a valence-bond treatment with localized three-centre bonds (see Chapter 7).

These two structures provide a reminder that many binary metal carbides M_xC_y contain discrete carbon atoms with coordination numbers higher than four. Aluminium carbide, Al_4C_3, for example, contains four-coordinate aluminium and both five- and six-coordinate carbon atoms. The bonding in such a compound must necessarily have a high degree of covalent character—a description in terms of adjacent discrete metal cations Al^{3+} and carbide anions C^{4-} is clearly unrealistic. Interestingly, aluminium carbide Al_4C_3 and aluminium nitride AlN are limiting members of a series of structurally related compounds $Al_xC_yN_z$ which contain invariably four-coordinate aluminium and nitrogen atoms, but five- or six-coordinate carbon atoms. In these systems, the Al–N bonding can be regarded as two-centre electron-pair bonding, whereas the Al–C bonding is of a lower order.[41]

References

1. Mackay, K. M., *Hydrogen Compounds of the Metallic Elements*, E. and F. N. Spon Ltd., London, 1966.
2. Turley, J. W., and H. W. Rinn, *Inorg. Chem.*, 1969, **8**, 18.
3. Breisacher, P., and B. Siegel, *J. Amer. Chem. Soc.*, 1964, **86**, 5053; 1965, **87**, 4255.
4. Young, A. R. jun., and R. Ehrlich, *J. Amer. Chem. Soc.*, 1964, **86**, 5359.
5. Greenwood, N. N., Gallium hydride and its derivatives, ch. 3 (pp. 37–64) of *New Pathways in Inorganic Chemistry*, Ebsworth, E. A. V., A. G. Maddock, and A. G. Sharpe (eds.), Cambridge Univ. Press, Cambridge, 1968.
6. Coates, G. E., and F. Glockling, *J. Chem. Soc.*, 1954, 22; Banford, L., and G. E. Coates, *J. Chem. Soc. (A)*, 1966, 274.
7. James, B. D., and M. G. H. Wallbridge, Metal tetrahydroborates, pp. 99–231 of *Progress in Inorganic Chem.*, vol. II (ed.), S. J. Lippard, Wiley, London, 1970.
8. Cook, T. H., and G. L. Morgan, *J. Amer. Chem. Soc.*, 1969, **91**, 774.
9. Sklar, N., and B. Post, *Inorg. Chem.*, 1967, **6**, 669.
10. Churchill, M. R., and R. Mason, The structural chemistry of organotransition metal complexes: some recent developments. *Advances in Organometal. Chem.*, 1967, **5**, 93–135.
11. Churchill, M. R., J. Wormwald, J. Knight, and M. J. Mays, *Chem. Comm.*, 1970, 458.
12. Vranka, R. G., and E. L. Amma, *J. Amer. Chem. Soc.*, 1967, **89**, 3121.
13. Byram, S. K., J. K. Fawcett, S. C. Nyburg, and R. J. O'Brien, *Chem. Comm.*, 1970, 16.
14. Dewar, M. J. S., and D. B. Patterson, *Chem. Comm.*, 1970, 544.

15. Magnuson, V. R., and G. D. Stucky, *J. Amer. Chem. Soc.,* 1969, **91**, 2544.
16. Malone, J. F., and W. S. McDonald, *Chem. Comm.,* 1967, 444; 1970, 280.
17. *Interatomic distances*, Special Publication No. 11, The Chemical Society, London, 1958.
18. See, for example, Jeffery, E. A., T. Mole, and J. K. Saunders, *Austral. J. Chem.,* 1968, **21**, 137, 649; Fishwick, M., C. A. Smith, and M. G. H. Wallbridge, *J. Organometal Chem.,* 1970, **21**, P9.
19. Cotton, F. A., and J. R. Leto, *J. Chem. Phys.,* 1959, **30**, 993.
20. Almenningen, A., S. Halvorsen, and A. Haaland, *Chem. Comm.,* 1969, 644.
21. Wagman, D. D., W. H. Evans, V. B. Parker, I. Halow, S. M. Bailey, and R. M. Schumm, Selected values of chemical thermodynamic properties, U.S. National Bureau of standards Technical Note 270-3, Jan. 1968.
22. Köster, R., and P. Binger, Organo-aluminium compounds, *Adv. Inorg. Chem. Radiochem.,* 1965, **7**, 263–348.
23. Reinheckel, H., K. Haage, and D. Jahnke, Organo-aluminium compounds in organic chemical reactions, *Organometal Chem. Reviews A,* 1969, **4**, 47–136.
24. Coates, G. E., and K. Wade, *Organometallic Compounds; 3rd Edn. Vol I; The Main Group Elements,* Methuen, London, 1967; Coates, G. E., M. L. H. Green, P. Powell, and K. Wade, *Principles of organometallic chemistry,* Methuen, London, 1968.
25. Ashby, E. C., J. Laemmle, and H. M. Neumann, *J. Amer. Chem. Soc.,* 1968, **90**, 5179.
26. Oliver, J. P., Fast exchange reactions of Group I, II, and III organometallic compounds, *Advances in organometal Chem.,* 1970, **8**, 167–209.
27. Ham, N. S., E. A. Jeffery, and T. Mole, *Austral. J. Chem.,* 1968, **21**, 2687.
28. Hata, G., *Chem. Comm.,* 1968, 7.
29. Amma, E. L., and R. E. Rundle, *J. Amer. Chem. Soc.,* 1958, **80**, 4141.
30. Sheldrick, G. M., and W. S. Sheldrick, *J. Chem. Soc. (A),* 1970, 28.
31. Malone, J. F., and W. S. McDonald, *J. Chem. Soc. (A),* 1970, 3362.
32. Snow, A. I., and R. E. Rundle, *Acta Cryst.,* 1951, **4**, 348.
33. Weiss, E., *J. Organometal Chem.,* 1964, **2**, 314.
34. Atwood, J. L., and G. D. Stucky, *J. Amer. Chem. Soc.,* 1969, **91**, 2538.
35. Brown, T. L., The structures of organo-lithium compounds, *Advances in Organometal Chem.,* 1965, **3**, 365–395.
36. Weiss, E., and E. A. C. Lucken, *J. Organometal Chem.,* 1964, **2**, 197; Weiss, E., and G. Hencken, *ibid.,* 1970, **21**, 265.
37. Weiss, E., and G. Sauermann, *J. Organometal Chem.,* 1970, **21**, 1.
38. Weiss, E., and G. Sauermann, *Angew. Chem., Internat. Edn.,* 1968, **7**, 133.
39. Braye, E. H., L. F. Dahl, W. Hübel, and D. L. Wampler, *J. Amer. Chem. Soc.,* 1962, **84**, 4633.
40. Sirigu, A., M. Bianchi, and E. Benedetti, *Chem. Comm.,* 1969, 596.
41. Jeffrey, G. A., and V. Y. Wu, *Acta Cryst.,* 1966, **20**, 538.

Problems

5–1   Describe, with examples, the types of compound that contain three-centre two-electron hydrogen bridges between two atoms.

5–2   Describe for associated metal alkyls $(MR_n)_x$ how the number and type of their alkyl bridges varies according to the group of the periodic table to which the metal M belongs.

5–3   The enthalpy of formation of aluminium tris(tetrahydroborate) ΔH_f° (at 25°) is $+13$ kJ/mole. Calculate the enthalpy of atomization of this compound, and hence, assuming a bond energy term E(B–H) of 375 kJ/mole for its terminal B–H bonds, calculate the bond energy term E(AlHB) for its AlHB bridge bonds. Comment on the relative values of E(M–H) and E(M–H–M′), where M and M′ are either aluminium or boron.

(Use the following enthalpies of atomization of the elements: Al, 326; B, 563; and H, 218 kJ/g atom.)

5-4 Explain what products may be expected from reactions between the following:

 (a) trimethylaluminium (2 mol) and triphenylaluminium (1 mol);

 (b) dimethylaluminium chloride (1 mol) and dimethylaluminium bromide (1 mol);

 (c) trimethylaluminium and an excess of diborane;

 (d) trimethylaluminium and an excess of water;

 (e) trimethylaluminium dimer (1 mol) and *t*-butyl alcohol (1 mol);

 (f) trimethylaluminium dimer (1 mol) and acetone (1 mol).

For those reactions which afford products with methyl groups still attached to aluminium, explain the number and relative intensities of the resonances due to methyl-protons expected in their ^1H n.m.r. spectra.

5-5 Compare the reactions of triethylaluminium with those of diborane.

5-6 Explain why the enthalpy of dissociation of dimeric aluminium compounds $(AlX_3)_2$ into planar monomers AlX_3 may not be a reliable guide to the relative strengths of the Al–X–Al bridge links in the dimers.

5-7 The enthalpy changes ΔH for the gas phase reaction:

$$Me_2S,MX_3 \rightarrow Me_2S + 1/n(MX_3)_n$$

where $MX_3 = BH_3$, $AlMe_3$, or $GaMe_3$ are 22, 36, and 34 kJ/mole respectively.

 If it is assumed that the reorganization energies of Me_2S and $GaMe_3$ are zero, what bond energy terms $E(M \leftarrow S)$ do these enthalpies correspond to?

5-8 The enthalpies of dissociation, ΔH, of dimeric aluminium alkyls $(R_3Al)_2$ vary with R as follows: R = Me, 83; R = Et, 71; R = Bu^i, 32 kJ/mole. Comment on the significance of these figures.

6 Carboranes

6-1 Introduction

Boron forms two major types of mixed hydride with carbon. In the first type, some examples of which have already been mentioned, typical hydrocarbon residues are attached to typical boron hydride residues by what may be regarded as localized two-centre electron-pair bonds. Examples of such organoboranes include the methyldiboranes $Me_nB_2H_{6-n}$ (where n can equal 1, 2, 3, or 4 but significantly *not* 5 or 6) and similar derivatives of higher boranes in which the organic groups replace terminal hydrogens of the parent boranes. In none of these does the organic group form an electron deficient bridge between two (or more) boron atoms of the type that is common among alkyl derivatives of the neighbouring more electropositive elements beryllium, magnesium, and aluminium.

In the second type of mixed hydride, the carboranes, which form the subject of this chapter, both boron and carbon atoms feature in the electron deficient molecular skeleton. Typically, these compounds have structures based on a complete or nearly complete triangular-faced cage of skeletal atoms, with normally one terminal hydrogen atom attached to each carbon and boron, and bridging hydrogen atoms located around the opening if the cage is incomplete. Their chemistry really developed only during the nineteen-sixties after their discovery among the products of reactions between acetylenes and boron hydrides.[1-3] Enough examples are now known for the main structural features and types of reaction of these compounds to be discerned. Structurally, they provide striking series of examples of the workings of the isoelectronic principle; they afford unique scope for the study of three-dimensional, delocalized electron systems; the closed cage neutral species are far more thermally stable and far less reactive than neutral boron hydrides of comparable molecular weight, so much so in fact that icosahedral carborane units have already found application in the backbones of thermally stable polymers,[4] where they are preferred to conventional organic links; moreover carborane anions with incomplete cage structures form transition metal complexes like metallocenes that underline the relationships between the carborane residues and aromatic hydrocarbons, and between metallocenes and electron deficient systems.

The most important series of carboranes is that having molecular formulae $C_2B_{n-2}H_n$, the dicarba-*closo*-boranes, where n is any integer from 5 to 12. These neutral species are isoelectronic with the borane anions $B_nH_n^{2-}$, which as was noted in Chapter 3 have closed shell electronic

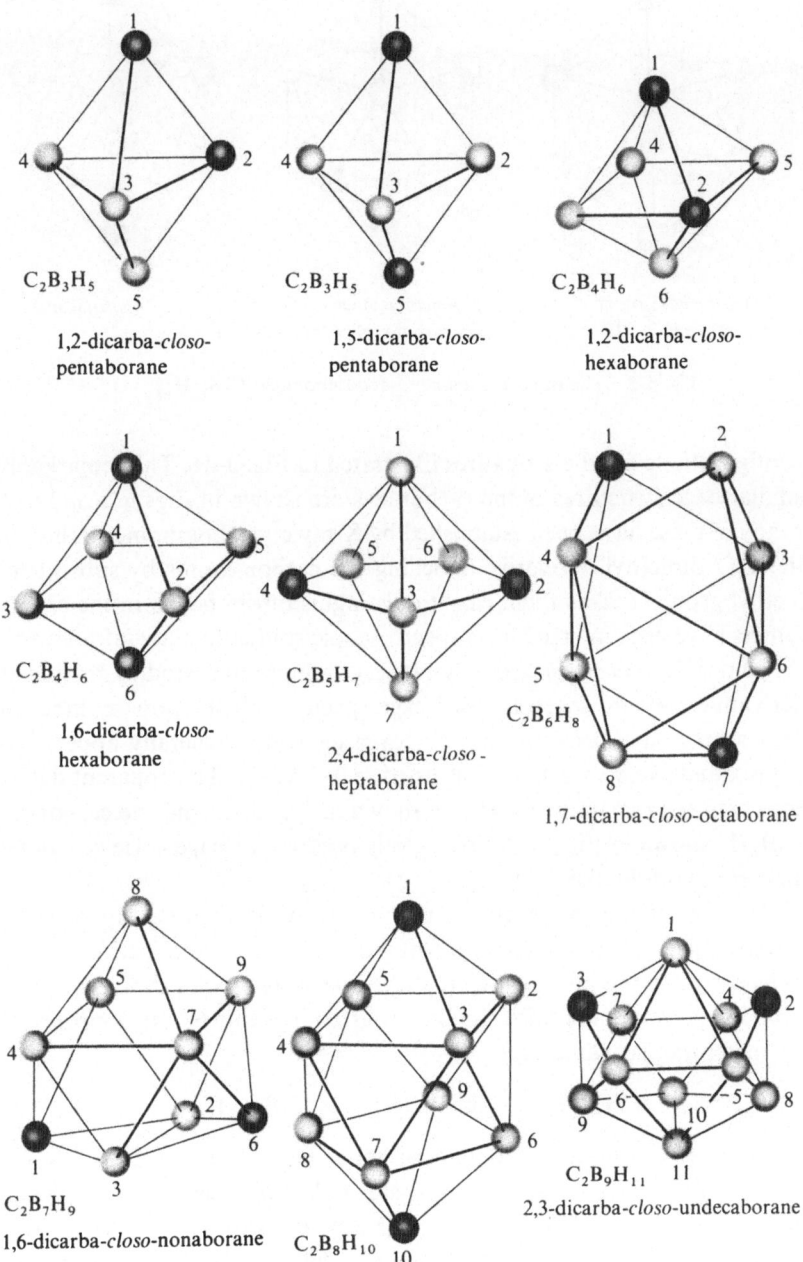

C₂B₃H₅
1,2-dicarba-*closo*-pentaborane

C₂B₃H₅
1,5-dicarba-*closo*-pentaborane

C₂B₄H₆
1,2-dicarba-*closo*-hexaborane

C₂B₄H₆
1,6-dicarba-*closo*-hexaborane

C₂B₅H₇
2,4-dicarba-*closo*-heptaborane

C₂B₆H₈
1,7-dicarba-*closo*-octaborane

C₂B₇H₉
1,6-dicarba-*closo*-nonaborane

C₂B₈H₁₀
1,10-dicarba-*closo*-decaborane

C₂B₉H₁₁
2,3-dicarba-*closo*-undecaborane

Fig. 6–1* Dicarba-*closo*-borane skeletons with from 5 to 11 cage atoms

*Terminally-attached atoms or groups are omitted from all the figures in this chapter.

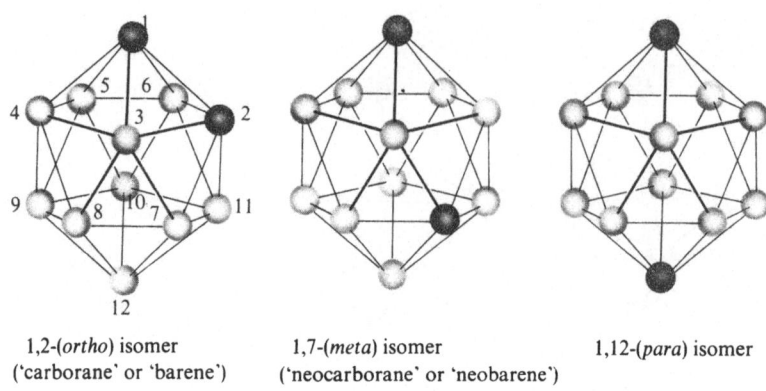

1,2-(*ortho*) isomer   1,7-(*meta*) isomer   1,12-(*para*) isomer
('carborane' or 'barene')   ('neocarborane' or 'neobarene')

Fig. 6–2 The three dicarba-*closo*-dodecaboranes, $C_2B_{10}H_{12}$

configurations for the structures illustrated in Fig. 3–16. The remarkably similar cage structures of the carboranes are shown in Figs 6–1 and 6–2. Many of these have been established by X-ray crystallographic studies on the C,C′ dimethyl derivative (labelling the carbon atoms by substituent methyl groups makes them easy to distinguish from boron in the cages). Others have been inferred from spectroscopic studies (mainly infrared and ^1H and ^{11}B n.m.r.) or from X-ray crystallographic studies on metal derivatives. Where several closed cage species with the same number of cage atoms have been studied, the cage geometry generally appears to approximate very closely to that listed in Table 6–1. The apparent difference between the anion $B_8H_8^{2-}$ as shown in Fig. 3–20 and the carborane $C_2B_6H_8$ shown in Fig. 6–1 arises largely because the cage is viewed along different two-fold axes.

In view of the isostructural and isoelectronic relationship between anions $B_nH_n^{2-}$ and neutral carboranes $C_2B_{n-2}H_n$, application of the isoelectronic principle suggests that the following species should all have closed cage structures, with the shapes listed in Table 6–1 as appropriate for the particular value of n:

$$B_nH_n^{2-} \qquad CB_{n-1}H_n^- \qquad C_2B_{n-2}H_n \qquad C_3B_{n-3}H_n^+$$

Though few carboranes containing three carbon atoms have as yet been made, and cationic species are unknown, several closed-cage monocarba anions $CB_{n-1}H_n^-$ are already known, as well as neutral compounds $CB_{n-1}H_{n+1}$ which have one bridging hydrogen and may be regarded as formally derived from the anions $CB_{n-1}H_n^-$ by addition of a proton. Where their structures are known, they have the expected closed cage geometry appropriate for species with n cage atoms and $(n+1)$ pairs of cage bonding electrons.

Another important series of carboranes consists of doubly charged

Table 6–1   Borane and carborane structural types

Number of skeletal bonding electron-pairs $(n+1)$	n-cornered polyhedron on which skeleton is based	closo- species $B_nH_n^{2-}$ $C_2B_{n-2}H_n$, etc.	nido- species $B_nH_n^{4-}$ $C_2B_{n-2}H_n^{2-}$, etc.	arachno- species $B_nH_n^{6-}$ $C_2B_{n-2}H_n^{4-}$, etc.
6	trigonal bipyramid	$C_2B_3H_5$		$B_3H_8^-$
7	octahedron	CB_5H_7, $CB_5H_6^-$ $C_2B_4H_6$, $B_6H_6^{2-}$	$C_2B_3H_7$ B_5H_9	B_4H_{10}
8	pentagonal bipyramid	$C_2B_5H_7$ $B_7H_7^{2-}$	B_6H_{10}, CB_5H_9 $C_2B_4H_8$, $C_2B_4H_6^{2-}$ $C_3B_3H_7$, $C_3B_3H_6^-$ $C_4B_2H_6$	B_5H_{11} $C_2B_3H_7^{2-}$
9	dodecahedron	$C_2B_6H_8$ $B_8H_8^{2-}$		
10	tricapped trigonal prism	$C_2B_7H_9$ $B_9H_9^{2-}$	$C_2B_6H_8^{2-}$	
11	bicapped Archimedean antiprism	$C_2B_8H_{10}$ $CB_9H_{10}^-$ $B_{10}H_{10}^{2-}$	$C_2B_7H_9^{2-}$	$C_2B_6H_8^{4-}$
12	octadecahedron	$C_2B_9H_{11}$, $CB_{10}H_{11}^-$ $B_{11}H_{11}^{2-}$	$B_{10}H_{14}$ $C_4B_6H_{10}$	$C_2B_7H_{13}$
13	icosahedron	$C_2B_{10}H_{12}$ $CB_{11}H_{12}^-$ $B_{12}H_{12}^{2-}$	$CB_{10}H_{14}$, $CB_{10}H_{13}^-$ $CB_{10}H_{11}^{3-}$, $C_2B_9H_{13}$ $C_2B_9H_{11}^{2-}$, $C_4B_7H_{11}$	$C_2B_8H_{10}^{4-}$

*Idealized for the anions $B_nH_n^{2-}$, which in practice tend to be slightly distorted in crystals. The carboranes may have lower symmetry, depending on which corners are occupied by the carbon atoms.

anions $C_2B_{n-2}H_n^{2-}$ or substances formally derived from them. These have n cage atoms and $(n+2)$ pairs of cage bonding electrons, i.e. one pair too many for the closed cage structure but just the right number for a cage with $n+1$ corners. Accordingly, they adopt *nido* (Greek for 'nest') structures in which one cage corner is left vacant, though when they are obtained as metal salts the metal cation may occupy the vacant site in the crystal. Isoelectronic species which might all be expected to have *nido*- structures are as follows:

$$B_nH_n^{4-} \quad CB_{n-1}H_n^{3-} \quad C_2B_{n-2}H_n^{2-} \quad C_3B_{n-3}H_n^- \quad C_4B_{n-4}H_n$$

The $B_nH_n^{4-}$ anions, though unknown as such, are effectively represented in protonated form by such familiar neutral boron hydrides as penta-borane(9), B_5H_9 (octahedral cage with one corner vacant; see Fig. 3–14), hexaborane(10), B_6H_{10} (pentagonal bipyramid with one apical corner vacant; see Fig. 3–8), and decaborane(14), $B_{10}H_{14}$ (the octahedron of Fig. 3–16 with the top corner vacant; see Fig. 3–10), each of which contains four bridging hydrogens. Hexaborane(10) in fact is the limiting member of a series of five compounds containing six cage atoms and eight pairs of cage bonding electrons, the other members of which (see Fig. 6–3)

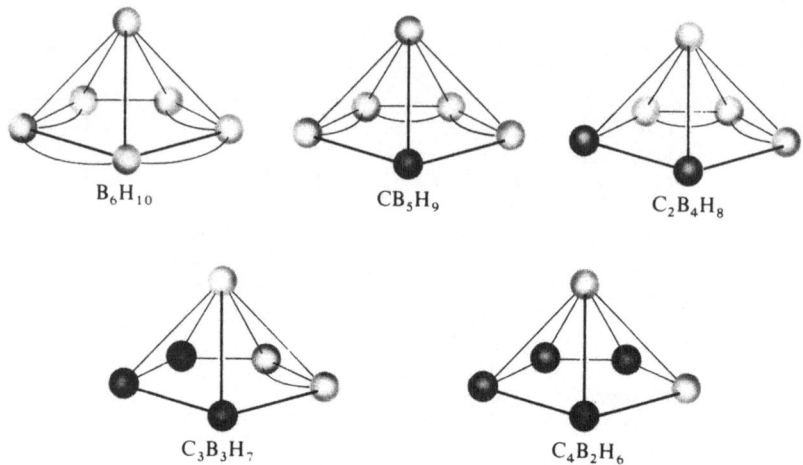

Fig. 6–3   The *nido* carboranes formally related to B_6H_{10}
All have eight pairs of electrons bonding the six cage atoms together.
Hydrogen bridges are represented by curved lines.

are the compounds monocarba-*nido*-hexaborane(9), CB_5H_9, dicarba-*nido*-hexaborane(8), $C_2B_4H_8$, tricarba-*nido*-hexaborane(7), $C_3B_3H_7$, and tetracarba-*nido*-hexaborane(6), $C_4B_2H_6$, the structures of all of which have a pentagonal pyramidal arrangement of the cage atoms. Those with bridging hydrogen atoms (all except $C_4B_2H_6$) have them linking basal boron atoms.

A third category of carborane, of which there are fewer known examples, is that series of compounds

$$B_nH_n^{6-} \quad CB_{n-1}H_n^{5-} \quad C_2B_{n-2}H_n^{4-} \quad C_3B_{n-3}H_n^{3-} \quad C_4B_{n-4}H_n^{2-}$$

$$C_5B_{n-5}H_n^{-}$$

for which the parent (hypothetical) anions $B_nH_n^{6-}$ are effectively the skeletons of boranes B_pH_{p+q} in which $q = 6$ (e.g., B_4H_{10} or B_5H_{11}). These have n cage atoms and $(n+3)$ pairs of cage bonding electrons, i.e., the right number for a cage with $(n+2)$ corners. They accordingly adopt *arachno*- (Greek for 'cobweb') structures in which two cage corners are left vacant. Examples include the neutral carborane $C_2B_7H_{13}$ (Fig. 6–5) and the anions $C_2B_8H_{10}^{4-}$ (Fig. 6–8), $B_3H_8^{-}$ (Fig. 3–13) and $C_2B_3H_7^{2-}$, believed to have a pentagonal shape like that of the cyclopentadienide anion $C_5H_5^{-}$, another member of the same *arachno*- series based on a pentagonal bipyramid.

Examples of the various categories of carborane and higher borane are listed in Table 6–1. It is clear that the number of pairs of skeletal bonding electrons is a key factor in the structures of these compounds, as this determines the polyhedron upon which the structure is based.[30,31]

When it is borne in mind that for many carboranes various isomers are possible, differing in the positions of the carbon atoms; that the interconversion of these isomers may involve yet further structures;[5,6] that the positions of the carbon atoms will influence the position of attack by nucleophiles or electrophiles; that methods are being devised for removing atoms from or adding atoms to the cages; that other elements than carbon and boron can be incorporated in these cages, it is seen that a vast chemistry of species related to cage boranes is only just being opened up.

In the following pages there is room merely to outline a few salient features of the chemistry of some representative species. More detailed treatments will be found among the reviews listed at the end of this chapter.[3–12,32,33]

6–2   The dicarba-*closo*-dodecaboranes, $C_2B_{10}H_{12}$

The dicarba-*closo*-dodecaboranes are the highest known members of the carborane series, and those whose chemistry has been most fully explored. Their 12 cage atoms adopt an icosahedral arrangement (see Fig. 6–2) in which opportunities for isomerism arise because the carbon atoms may occupy adjacent (1,2), alternate (1,7), or *trans* (1,12) positions. (The skeletal atoms of carborane frameworks are identified by the numbering systems illustrated in Figs 6–1 and 6–2; numbering begins at one of the apices, proceeds clockwise through the lower layers of boron atoms as viewed from above the apex, ending in the case of a bipyramidal structure at the other apex.) The three dicarba-*closo*-dodecaboranes are commonly referred to as '*ortho*-', '*meta*-' (or '*neo*'), and '*para*'-carborane respectively; the name 'barene' is used in the Russian literature.

The 1,2-isomer, and various C-substituted derivatives, are obtained by reactions between acetylenes and decaboranes in the presence of a Lewis base:

$$B_{10}H_{14} + R^1C\mathop{:}CR^2 \xrightarrow{L} R^1C \underset{B_{10}H_{10}}{\diagdown\!\!\bigcirc\!\!\diagup} CR^2 + 2H_2$$

The reactions probably involve the derivatives $B_{10}H_{12}L_2$ as intermediates. Best yields are obtained when L is a dialkyl sulphide, though nitriles, phosphines, and tertiary amines are also effective. The groups R can be alkyl, aryl, or hydrogen, or functional groups that do not contain reactive hydrogen.

The parent compound, $1,2\text{-}C_2B_{10}H_{12}$, is remarkably thermally stable, remaining unaffected by heat up to about 470° when it rearranges smoothly and in good yield into the *meta* isomer, $1,7\text{-}C_2B_{10}H_{12}$. This in turn gives the *para* isomer $1,12\text{-}C_2B_{10}H_{12}$, though in poor yield, at 615°. Extensive breakdown of the molecular skeleton takes place at about 630°. Deca-borane (14) by contrast decomposes at 250°.

It appears likely that these rearrangements occur via square-faced polyhedral intermediates (see Fig. 6–4).[5,6] A 'diamond-square-diamond'

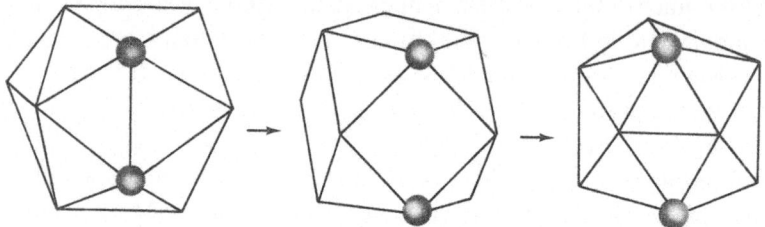

A possible route from $1,2\text{-}C_2B_{10}H_{12}$ to $1,7\text{-}C_2B_{10}H_{12}$ via an intermediate with cubo-octahedral geometry (an example of a 'diamond-square-diamond' mechanism)

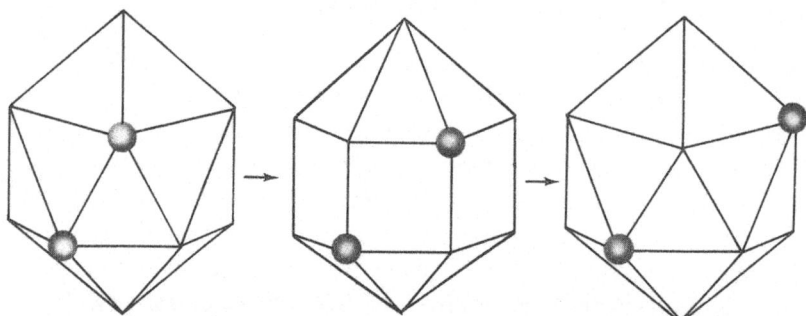

An alternative route from $1,2\text{-}C_2B_{10}H_{12}$ to $1,7\text{-}C_2B_{10}H_{12}$ by mutual rotation of the two pentagonal pyramidal halves of the molecule.

Fig. 6–4   Possible mechanisms of rearrangement of dicarba-*closo*-dodecaboranes.

rearrangement involving a cubo-octahedral intermediate would readily account for the *ortho* → *meta* isomerization, but would not allow con-version of the *meta* isomer into the *para* (which certainly requires more forcing conditions) unless the triangular faces of the cubo-octahedron were able to rotate.[13] Such a local rotation of a set of atoms relative to the rest of the molecule appears more likely than a mechanism involving mutual rotation of the two pentagonal pyramidal halves of the molecule, which would be equally applicable to both *ortho* → *meta* and *meta* → *para* rearrangements. The direction of these rearrangements is normally that to increase the separation of the carbon atoms, favouring ultimate forma-tion of the thermodynamically most stable isomer, *para*-carborane, $1,12\text{-}C_2B_{10}H_{12}$.

In marked contrast to the boron hydrides B_pH_{p+q}, the carboranes are not sensitive to hydrolysis, being unaffected by boiling methanol or water. The carborane skeleton is not affected by such strong oxidants as permanganate, hypochlorite, and hydrogen peroxide or by reducing agents. It is, however, susceptible to attack by nucleophiles such as methoxide ion or hydrazine, which abstract a boron atom from the cage and lead to the *nido*-anions $C_2B_9H_{12}^-$ and $C_2B_9H_{11}^{2-}$ (see below). Apart from this degradation of the cage by bases, the icosahedral carboranes have an extensive derivative chemistry.[7]

Photochemical chlorination of *ortho*-carborane with gaseous chlorine effects stepwise substitution at the boron atoms, leading ultimately to the decachloro derivative $C_2H_2B_{10}Cl_{10}$, which is an acid of comparable strength to carboxylic acids. Its C–H bonds are chlorinated only under very severe conditions. The order in which the B–H bonds suffer *electrophilic* chlorination or bromination in *ortho*-carborane is as follows:

$$9,12 > 8,10 > 4,5,7,11 > 3,6 > 1,2$$

This sequence is believed to reflect the decreasingly negative charge on the atoms in the same order; the boron atoms remote from the two carbons bear the greatest negative charge; those next to both carbons are relatively positively charged, as are the carbon atoms. This is intelligible in that, although carbon is more electronegative than boron, it provides one electron more than does boron to the cage bonding and so suffers greater drainage of electronic charge into the cage.

The weakly acidic carbon-attached hydrogen atoms of *ortho*-carborane can be replaced by lithium atoms by reaction with lithium alkyls; the products are useful intermediates from which to effect substitution at carbon, exhibiting the usual reactivity of organometallic reagents, for example,

*N-bromosuccinimide

The benzocarborane formed in this second reaction sequence provided the first clear evidence of conjugation of the 'aromatic' carborane nucleus with an organic π-bonded system. Both the ^1H n.m.r. spectrum (a multiplet centred at $3.63\ \tau$ between typical olefinic and aromatic values) and the ultraviolet spectrum (similar to those of other benzo- derivatives) indicate that the six-membered carbon ring has considerable aromatic character.

Carboranyl Grignard reagents can be prepared using standard procedures, and also by rearrangement of 1-carboranyl-methylmagnesium bromide:

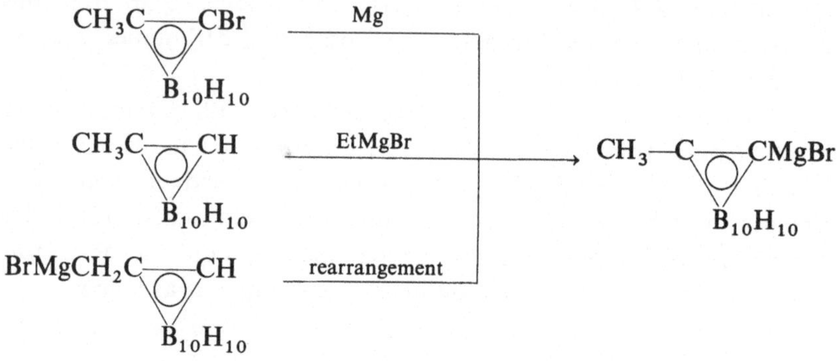

These C-metalated compounds have been used for the preparation of a wide range of functionally substituted carboranes. Much attention has been paid to the preparation of difunctional derivatives from which polymers might be prepared, for example,[4]

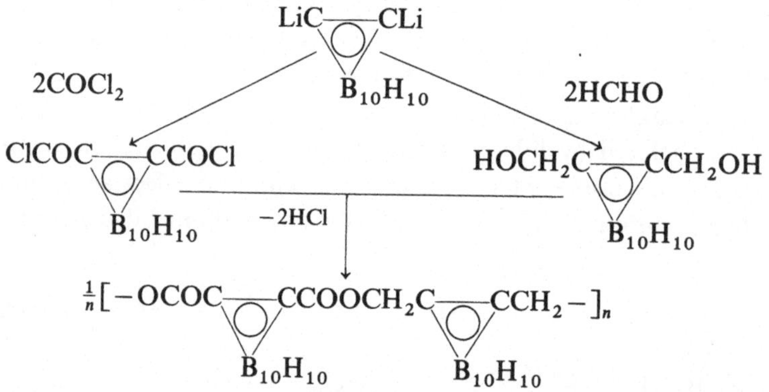

Derivatives of *ortho*-carborane are not very convenient for this purpose, because they tend to give cyclic monomers or oligomers rather than polymers:

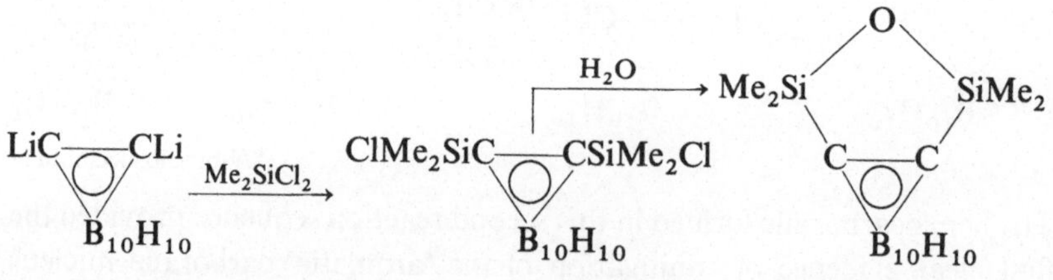

Polymers are obtained, however, from derivatives of *meta*-carborane:

$$\xrightarrow[-2MeCl]{ClMe_2SiCB_{10}H_{10}CSiMe_2Cl} \frac{2}{n}\left[-\underset{Me}{\overset{Me}{O Si}}CB_{10}H_{10}C\underset{Me}{\overset{Me}{Si}}-\right]_n$$

$$\underset{Me\quad Me}{MeOSiCB_{10}H_{10}CSiOMe}$$

$$\xrightarrow[-2MeCl]{ClMe_2SiOSiMe_2Cl} \frac{1}{n}\left[-\underset{Me}{\overset{Me}{O Si}}CB_{10}H_{10}C\underset{Me\ Me\ Me}{\overset{Me\ Me\ Me}{Si O Si O Si}}-\right]_n$$

The products vary according to the proportions of siloxane and silyl-carborane in the polymer, from highly crystalline solids to elastomeric gums. They decompose at about 330°C in air (when incidentally the *methyl* groups start to suffer oxidation), and at 450° under nitrogen, and appear likely to find application where heat-resistant properties are particularly important.[4]

High thermal stability also characterizes carboranyl links to other metalloids and less electropositive metals, links which are readily generated by reactions between the lithio derivative and an appropriate halide:

$$R\text{-}C\underset{B_{10}H_{10}}{\diagup}\text{-}CLi \quad + HgCl_2 \rightarrow \quad R\text{-}C\underset{B_{10}H_{10}}{\diagup}\text{-}C\text{-}Hg\text{-}C\underset{B_{10}H_{10}}{\diagup}\text{-}C\text{-}R$$

The mercury compounds $(RC_2B_{10}H_{10})_2Hg$ for example are stable to above 350°, in contrast to simple mercury alkyls, which readily decompose to form alkyl radicals and/or olefins. The reaction

$$Ph\text{-}C\underset{B_{10}H_{10}}{\diagup}\text{-}C\text{-}HgMe \quad \xrightarrow[EtOH]{HCl} \quad Ph\text{-}C\underset{B_{10}H_{10}}{\diagup}\text{-}CH \quad + MeHgCl$$

illustrates the electron withdrawing properties of the carboranyl group, which are similar to those of pentafluorophenyl or pentachlorophenyl groups.

Substitution of both carbon atoms of *ortho*-carborane by amino, phosphino, arsino, or mercapto groups affords species capable of acting as bidentate ligands:

$$B_{10}H_{10}\diagup\diagup\underset{C-P}{\overset{C-P}{\underset{Ph_2}{\overset{Ph_2}{|}}}} \quad + Fe(CO)_5 \xrightarrow{150°} B_{10}H_{10}\diagup\diagup\underset{C-P}{\overset{C-P}{\underset{Ph_2}{\overset{Ph_2}{|}}}}Fe(CO)_3$$

Particular attention has been paid to transition metal complexes in which the donor groups are of the 'π-acid ligand' type (R_2P, R_2As, RS).[14] Interestingly, in these the carborane residue does *not* appear to provide an effective network for π-delocalization; the π-acidity of such ligands appears less than that of their *ortho*-phenylene counterparts.

6–3   Other dicarba-*closo*-boranes

Carboranes containing from six to nine boron atoms can be prepared by a series of degradation reactions starting from icosahedral species:[15]

$$C_2B_{10}H_{12} \xrightarrow[\text{2. H}^+]{\text{1. OR}^-} C_2B_9H_{13} \xrightarrow{130°} C_2B_9H_{11} \xrightarrow[\text{aq. HOAC}]{Cr_2O_7^{2-}} C_2B_7H_{13.} \xrightarrow{200°}$$

$$C_2B_{n-2}H_n$$

$$(n = 8, 9, \text{ or } 10)$$

In this sequence, the first step involves removal from the icosahedron of one of the two boron atoms that are linked to both carbons. This is believed to occur by way of the following sequence of reactions:

$$C_2B_{10}H_{12} \xrightarrow{\text{OMe}^-} [C_2B_{10}H_{12}OMe]^- \xrightarrow{\text{OMe}^-} [C_2B_{10}H_{12}(OMe)_2]^{2-}$$

$$C_2B_9H_{12}^- + H_2 + B(OMe)_3 \xleftarrow{\text{MeOH}} C_2B_9H_{11}^{2-} + HB(OMe)_2$$

That the 3 (or 6) boron atom is removed is intelligible in that these are the most positively charged boron atoms in the cage and so most susceptible to nucleophilic attack by methoxide ions. The product after acidification, $C_2B_9H_{13}$, has the *nido* structure shown in Fig. 6–5 in which the two carbon atoms define one edge of the open pentagonal face, and the

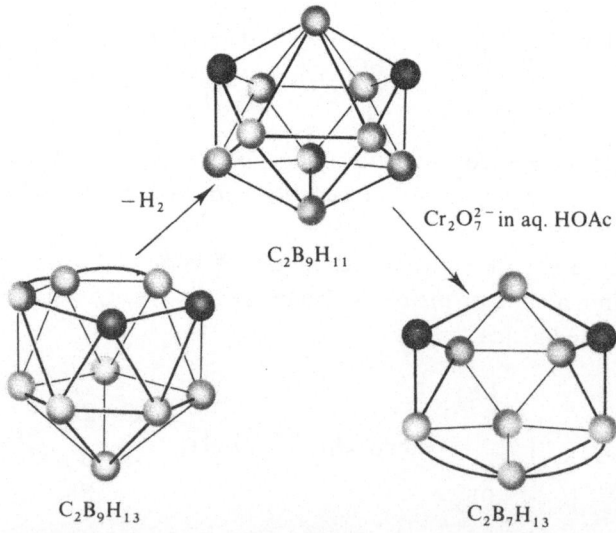

$-H_2$ $C_2B_9H_{11}$ $Cr_2O_7^{2-}$ in aq. HOAc

$C_2B_9H_{13}$ $C_2B_7H_{13}$

Fig. 6–5   Key species in the preparation of certain lower carboranes. (The carbon atoms of $C_2B_7H_{13}$ each have two terminal hydrogens attached; all other cage atoms have one terminal hydrogen. Curved lines represent hydrogen bridges.)

bridging hydrogen atoms link the three boron atoms which complete the pentagonal face. Typically for a *nido* carborane containing bridging hydrogen atoms, this compound decomposes when heated, losing hydrogen to form the *closo* dicarba-undecaborane, $C_2B_9H_{11}$ (Fig. 6–5). The octadecahedral structure of this compound has been established by a crystallographic study of the C,C′-dimethyl derivative.

Degradation of this dicarba-undecaborane is effected by the action of dichromate ion in aqueous acetic acid, which interestingly affords a product, $C_2B_7H_{13}$, with two boron atoms fewer, but two hydrogen atoms more, than the starting material. The structure of dicarba-*arachno*-nonaborane (13), $C_2B_7H_{13}$, which contains two methylene groups linked in each case to three boron atoms, is shown in Fig. 6–5. Note its relationship to the 11-atom cage of $C_2B_9H_{11}$ also shown in Fig. 6–5; formally, as a protonated derivative of the hypothetical anion $C_2B_7H_9^{4-}$, it would be expected to adopt a structure based on that of the 11-atom cage species but with *two* corners vacant, since it contains *two* pairs of electrons more than the number appropriate for a nine-atom closed cage species.

The *closo* carboranes with six, seven, or eight boron atoms are formed by the thermal decomposition of $C_2B_7H_{13}$, which involves disproportionation as well as hydrogen elimination:

$$C_2B_7H_{13} \xrightarrow[Ph_2O]{200°} 30\% \ 1,7\text{-}C_2B_6H_8 + 7\% \ 4,5\text{-}C_2B_7H_9 + 30\% \ 1,6\text{-}C_2B_8H_{10}$$

In reactions such as this which lead to mixtures of novel products, some in very low yield, mass spectroscopy can be extremely useful in their identification,[3] particularly since chemical analysis of such boron–carbon systems is difficult. The presence of both ^{10}B and ^{11}B isotopes in relative abundance, ca 1:4, causes ions containing several boron atoms to appear on mass spectra as sets of peaks, the number and relative intensities of which are characteristic of the number of boron atoms. Careful analysis of the isotope pattern accordingly allows the number of boron atoms in each fragment to be deduced and empirical formulae to be obtained.

The carboranes with still fewer boron atoms are prepared by similar thermal or photolytic decomposition of a *nido*-intermediate, in this case $C_2B_4H_8$ (Fig. 6–3) which itself is best prepared from pentaborane(9) and acetylene:[16]

$$B_5H_9 + HC\!:\!CH \xrightarrow[48h]{215°} 2,3\text{-}C_2B_4H_8$$

$$\begin{array}{l} \xrightarrow[23°C/2h]{u.v.} C_2B_3H_5 + C_2B_4H_6 \text{ isomers} \\[2mm] \xrightarrow[450\text{-}460°]{10mm/1\text{-}3\ s.} C_2B_3H_5 + C_2B_4H_6 \text{ isomers} \\[1mm] \hspace{4cm} + 2,4\text{-}C_2B_5H_7 \end{array}$$

30–40% yield

The *nido*-intermediate $C_2B_4H_8$ need not be isolated if continuous flow methods at higher temperatures are used:

$$B_5H_9 + C_2H_2 + H_2 \xrightarrow[\substack{\text{nominal residence} \\ \text{time of } ca.\ 0\cdot5\ \text{s}}]{500°C} \text{50–60\% yield of the three}$$
$$1\quad:\quad 4\quad:\quad 40 \qquad \text{small } closo \text{ carboranes}$$

These small carboranes are relatively volatile substances; 1,5-dicarba-*closo*-pentaborane $C_2B_3H_5$ has m.p. $-126°$, b.p. $-3\cdot7°$, and 1,6-$C_2B_4H_6$ has m.p. $-32°$, b.p. $22\cdot7°C$. There is a surprising dearth of information on the melting and boiling points of several other parent species, but volatilities decrease with increasing molecular weight.

The dicarba-*closo*-boranes in general are relatively unreactive by comparison with the higher boron hydrides. They are air-stable, and do not react at room temperature with the sort of substances (acetone, water, amines) that typically attack substances with bridging hydrogen atoms or BH_2 groups. When heated, they normally isomerize to the isomer which is most thermally stable, which is usually that in which the carbon atoms are separated and occupying the positions of lowest coordination number. For example, a *trans* arrangement of the carbon atoms is found in the most stable isomers of $C_2B_3H_5$ and $C_2B_4H_6$ (Fig. 6–1), whereas in the case of the pentagonal bipyramidal dicarba-*closo*-heptaborane $C_2B_5H_7$ the most stable isomer is that (Fig. 6–1) in which the carbon atoms occupy the non-adjacent equatorial positions 2 and 4. As with icosahedral carboranes, interconversion of isomers is believed to occur via square-faced intermediates.

Isomers of the small *closo* carboranes are readily identified by their ^{11}B n.m.r. spectra, in which a doublet absorption is expected for each magnetically distinct set of boron atoms, since each boron has one terminal hydrogen atom attached. As the cages become more complicated, and the differences between individual boron atoms become less marked, the overlapping of signals makes identification of individual isomers more difficult for higher carboranes, although isomers of high symmetry may still be recognized from their ^{11}B n.m.r. spectra alone, as illustrated in Fig. 6–6. For example, the compounds, 1,5-$C_2B_3H_5$, 1,6-$C_2B_4H_6$, 1,10-$C_2B_8H_{10}$, and 1,12-$C_2B_{10}H_{12}$ all have very simple ^{11}B n.m.r. spectra, consisting in each case of a single doublet with $J(^{11}B—^1H)$ 183, 188, 162, and 160 Hz respectively and chemical shifts, δ, $-1\cdot4$, $+19\cdot4$, $+10\cdot3$, and $+15\cdot2$ ppm respectively relative to boron trifluoride-ether complex. The isomer, 1,2-$C_2B_4H_6$, however, has a spectrum consisting of two doublets of equal intensity attributable to the magnetically distinct pairs of boron atoms, numbers 3 (and 5) and 4 (and 6) (see Fig. 6–1). These two doublets overlap to give the appearance of a triplet (see Fig. 6–6) if the spectrum is recorded at a frequency of 12·8 MHz.

In the thermodynamically most stable isomer of dicarba-*closo*-hepta-

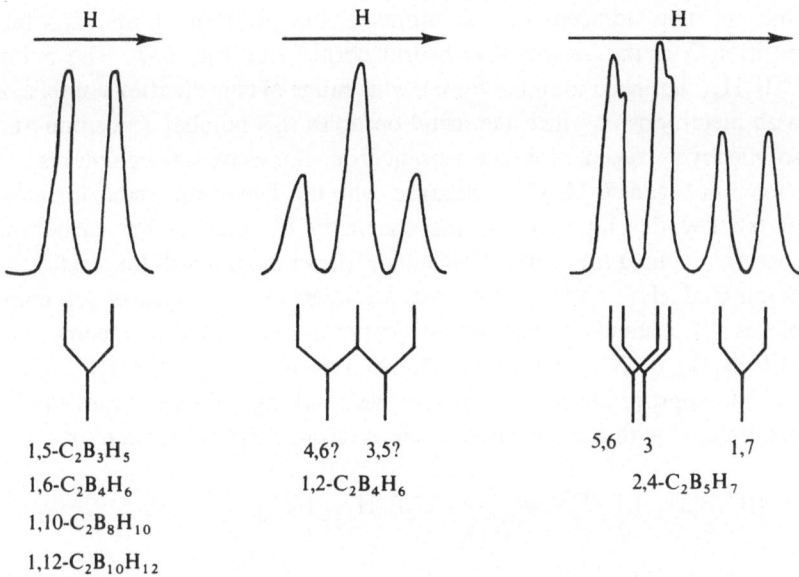

Fig. 6–6 ^{11}B n.m.r. spectra of some *closo*-carboranes, recorded at 12·8 MHz.

borane, $2,4\text{-}C_2B_5H_7$ (Fig. 6–1), there are three distinct types of boron atom; the unique boron (number 3), the apical pair (1 and 7), and the equatorial pair (5 and 6). These give rise to three doublets, of relative intensities $1:2:2$. The intense doublet located at high field, appropriate for relatively negatively charged boron atoms, is attributable to the apical borons. The remaining two doublets, which are nearly superimposed, are assignable to the equatorial borons (Fig. 6–6).

6–4   Dicarba-*nido*-borane anions and their metal complexes

The *nido* compound $C_2B_9H_{13}$, which has already been seen to be an important intermediate in the preparation of lower carboranes from dicarba-*closo*-dodecaborane $C_2B_{10}H_{12}$, can function as a dibasic acid by ionization of its two bridging hydrogen atoms (see Fig. 6–5) as protons. Salts of the anions $C_2B_9H_{12}^-$ and $C_2B_9H_{11}^{2-}$ can be prepared as follows:

$$1,2\text{-}C_2B_{10}H_{12} \xrightarrow[\text{MeOH}]{\text{OMe}^-} (3)\text{-}1,2\text{-}C_2B_9H_{12}^- \begin{cases} \xrightarrow{\text{Me}_3\text{NH}^+\text{Cl}^-} \text{Me}_3\text{NH}^+C_2B_9H_{12}^- \\ \\ \xrightarrow{\text{NaH/THF}} (\text{Na}^+)_2C_2B_9H_{11}^{2-} \end{cases}$$

The convention for numbering the cage atoms of these *nido*-anions is that the two adjacent carbon atoms define positions 1 and 2, while position 3 is the vacant icosahedral corner (see Fig. 6–7). The anion $C_2B_9H_{11}^{2-}$ has been found to form a wide range of coordination complexes with metal ions in which the metal occupies this number 3 position and so effectively regenerates the icosahedron. Some typical complexes are shown in Fig. 6–7. M. F. Hawthorne, who has been responsible for most of the key developments in the chemistry of these higher carborane species,[17] coined the word 'dicarbollide' (from the Spanish for jar) for the anion $C_2B_9H_{11}^{2-}$, and its complexes are referred to as 'dicarbollyl' complexes. The standard method of preparing bis-dicarbollyl complexes $M(C_2B_9H_{11})_2^{n-}$ is by the reaction between the disodium salt $Na_2C_2B_9H_{11}$ and the appropriate metal dihalide; the resulting anionic species can be precipitated with a large cation such as tetramethylammonium ion:

$$[(C_2B_9H_{11})_2Fe^{II}]^{2-} \underset{\substack{Na/Hg \\ in\ MeCN}}{\overset{O_2}{\rightleftharpoons}} [(C_2B_9H_{11})_2Fe^{III}]^-$$

$$FeCl_2\ in\ THF\ under\ N_2$$

$$Na_2C_2B_9H_{11} \xrightarrow[\substack{aqueous\ OH^-}]{Cu^{II}\ salt} [(C_2B_9H_{11})_2Cu^{II}]^{2-} \xrightarrow{air} [(C_2B_9H_{11})_2Cu^{III}]^-$$

$$Ni^{II}\ salt\ in\ THF\ in\ air$$

$$\xrightarrow{FeCl_3\ in\ H_2O} (C_2B_9H_{11})_2Ni^{IV}$$

$$[(C_2B_9H_{11})_2Ni^{III}]^-$$

$$\xrightarrow{reduction} [(C_2B_9H_{11})_2Ni^{II}]^{2-}$$

Mixed complexes such as the dicarbollyl(cyclopentadienyl)iron(III) species shown in Fig. 6–7 can be synthesized in an analogous manner:

$$Na_2C_2B_9H_{11} + NaC_5H_5 + FeCl_2 \xrightarrow{boiling\ THF} C_5H_5Fe^{III}(C_2B_9H_{11})$$

$$Na_2C_2B_9H_{11} + BrMn(CO)_5 \xrightarrow{boiling\ THF} Na^+[(C_2B_9H_{11})Mn(CO)_3]^-$$

The bis(dicarbollyl) complexes are far more thermally stable than their bis(cyclopentadienyl) analogues. The iron complexes $[(C_2B_9H_{11})_2Fe]^{2-}$ and $[(C_2B_9H_{11})_2Fe]^-$, which are formally analogous to ferrocene $(C_5H_5)_2Fe$ and ferricinium ion $[(C_5H_5)_2Fe]^+$ respectively, also exhibit much greater stability to acids; the iron(III) complex is unchanged by heating to 300° and does not react with strong acids, in marked contrast to ferricinium salts, which are relatively difficult to isolate and work with.

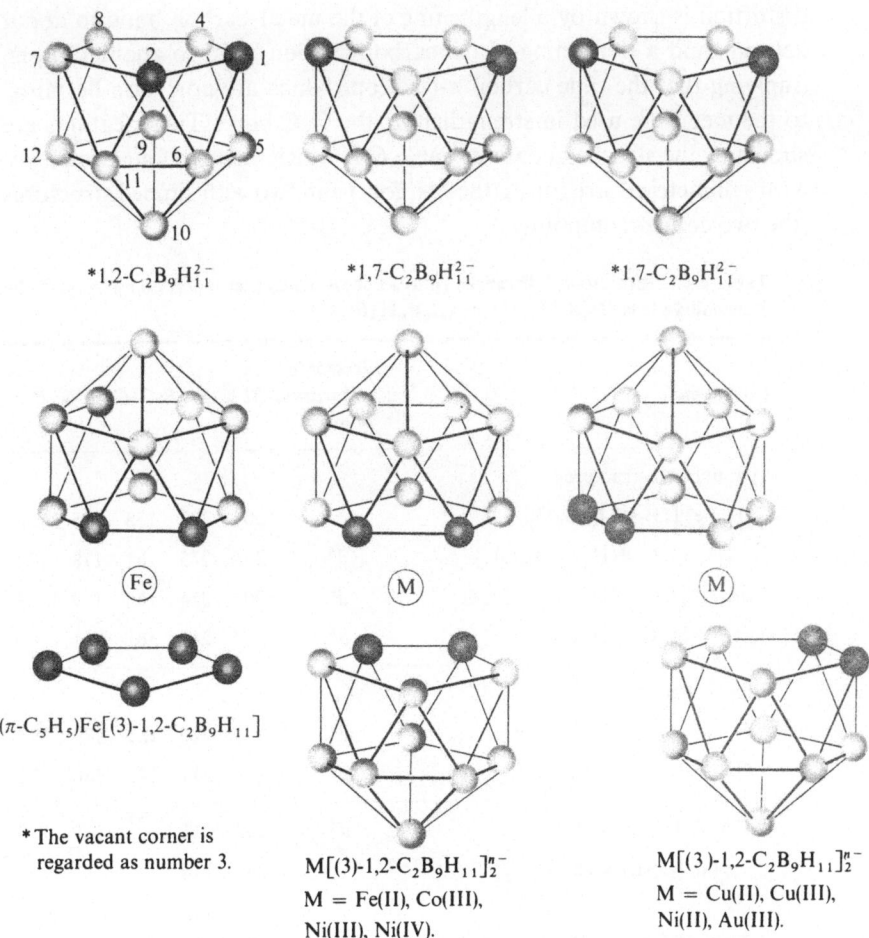

Fig. 6-7 Some *nido*-carborane anions with one icosahedral site vacant, and typical transition metal complexes

Several bis(dicarbollyl) complexes have been studied by X-ray crystallography,[18] and two structural types have been found, both of which are shown in Fig. 6–7. In one type, formed by metal ions with a d^5 (Fe(III)), d^6 (Fe(II)), Co(III), or Ni(IV)) or d^7 (Co(II) or Ni (III)) electronic configuration, the metal is located centrally over the pentagonal face of the $C_2B_9H_{11}^{2-}$ anion. The other type, formed by ions with a d^8 (Ni(II) or Cu(III)), or d^9 (Cu(II)) configuration, has a distorted or 'slipped' structure in which the metal atom is displaced towards the three boron atoms, and away from the two carbon atoms, of the pentagonal face, a distortion which has been interpreted as possibly indicative of metal–ligand interactions more akin to π-allyl than to π-cyclopentadienyl interactions, since in these 'slipped' complexes the metal has a primary coordination sphere of six boron atoms, three from each cage. This distortion is shown by a lengthening of the metal–carbon bond to about 250 pm, and a shortening of the carbon–carbon bond to about 150 pm, implying that the cage carbon π-electrons which are now 'non-bonding' to the metal are used in strengthening the C–C bond. These features are shown by the structural data in Table 6–2, which relate to four complexes with symmetrical structures (the first four) and two with slipped structures (the two copper compounds).

Table 6–2 Interatomic distances (pm) in some transition metal complexes of the dicarbollide ions $(C_2B_9H_{11})^{2-}$ or $(C_2B_9H_8Br_3)^{2-}$

Compound	Electronic configuration of metal	M–C	M–B	C–C	B–B	B–C
Symmetrical structures						
$C_5H_5Fe^{III}[(3)-1,2-C_2B_9H_{11}]$	d^5	204	209	158	175	168
$Cs^+[(3)-1,2-C_2B_9H_{11}Re^I(CO)_3]^-$	d^6	231	235	161	178	172
*$Me_4N^+[Co^{III}\{(3)-1,2-C_2B_9H_8Br_3\}]^-$	d^6	202	214	165	179	173
$Ni^{IV}[(3)-1,2-C_2B_9H_{11}]_2$	d^6	207	211	161		
Slipped structures						
†$(Et_4N^+)_2[Cu^{II}\{(3)-1,2-C_2B_9H_{11}\}_2]^{2-}$	d^9	257	220	153	179	172
‡$(Ph_3PMe)^+[Cu^{III}\{(3)-1,2-C_2B_9H_{11}\}_2]^-$	d^8	252	211	148	180	171
$[MeN(C_2H_4)_3NMe]^{2+}$	d^8	212	211	—	179	170
$[Ni^{II}\{(3)-1,7-C_2B_9H_{11}\}_2]^{2-}$		239	214			
			218			

Isomorphous species:
*$[Ni^{III}(C_2B_9H_{11})_2]^-$ †$[Ni^{II}(C_2B_9H_{11})_2]^{2-}$ ‡$[Au^{III}(C_2B_9H_{11})_2]^-$

The last compound listed in Table 6–2 is a complex of an isomeric ligand, $1,7$-$C_2B_9H_{11}^{2-}$ in which the carbon atoms still border the pentagonal face, but are no longer adjacent (see Fig. 6–7). In this complex, $[Ni(1,7-C_2B_9H_{11})_2]^{2-}$ the nickel(II) has a d^8 configuration, but there is little distortion from the symmetrical structure. The nickel atom is shifted off-centre by only 15 pm as compared with a slip of 60 pm for $[Cu(1,2-C_2B_9H_{11})_2]^-$ and $[Ni(1,2-C_2B_9H_{11})_2]^{2-}$. However, this complex is unusual in that the pentagonal face is non-planar; the carbon atoms are bent away from the nickel back into the cage. This non-planarity, plus the small but real slip distortion, causes one carbon atom to be much further from the metal than the remaining atoms of the pentagonal face (239 pm as compared with 214 pm average).

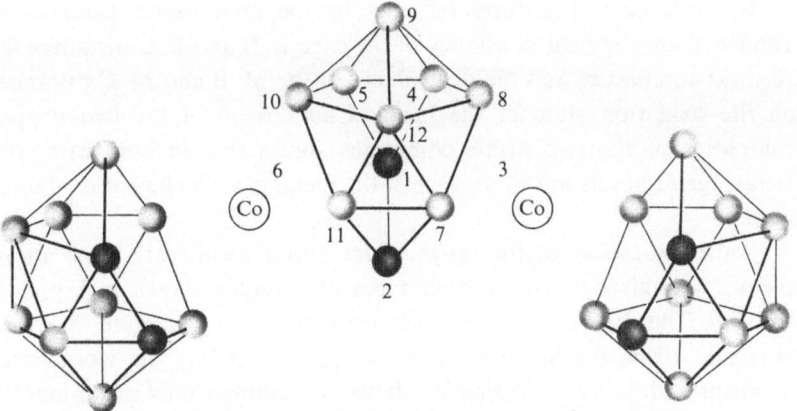

Fig. 6–8 The structure of the cobalt(III) complex, $[(C_2B_9H_{11})Co(C_2B_8H_{10})Co(C_2B_9H_{11})]^{2-}$, in which the central bridging unit, the 'dicarbacanastide' ion, $C_2B_8H_{10}^{4-}$, is a ten-particle icosahedral fragment with two non-adjacent sites (numbers 3 and 6) vacant.

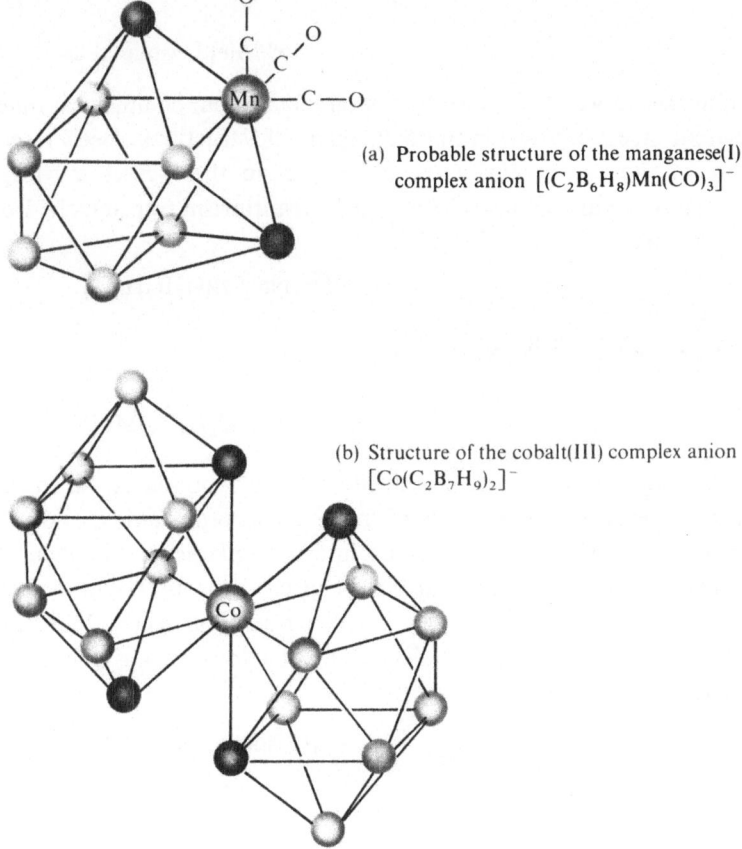

(a) Probable structure of the manganese(I) complex anion $[(C_2B_6H_8)Mn(CO)_3]^-$

(b) Structure of the cobalt(III) complex anion $[Co(C_2B_7H_9)_2]^-$

Fig. 6–9 Incorporation of transition metals in nine- and ten-atom polyhedra: complexes of the *nido*-anions (a) $C_2B_6H_8^{2-}$ and (b) $C_2B_7H_9^{2-}$

Other structural features reflected by the interatomic distances in Table 6–2 are the near constancy of the cage B–B and B–C distances for all these complexes, and the dependence of the M–B and M–C distances on the oxidation state of the metal. Comparison of the two copper complexes or the two nickel complexes shows that in both cases the metal–ligand bonds are shorter when the metal is in the higher oxidation state.

Some indication of the range of transition metal–carborane anion chemistry is given by the further types of complex shown in Figs 6–8 and 6–9. The first shows a complex containing the bidentate 'dicarbacanastide' (from the Spanish for 'basket') ion $C_2B_8H_{10}^{4-}$, an icosahedral fragment with two corner sites available for occupancy by metal ions;[19] this results from alkaline degradation of the dicarbollide ion:

$$[(C_2B_9H_{11})_2Co^{III}]^- \xrightarrow[\substack{+ \text{ excess CoCl}_2 \\ 100°}]{30\% \text{ NaOH}}$$

$$[(C_2B_9H_{11})Co^{III}(C_2B_8H_{10})Co^{III}(C_2B_9H_{11})]^{2-}$$

15% yield: isolated as Cs^+ salt.

Complexes in which a single transition metal atom completes a nine- or 10-atom cage are shown in Fig. 6–9. Again, the structures closely resemble those of the parent carborane polyhedra, in these cases a tricapped trigonal prism and bicapped Archimedean antiprism respectively. Routes to these species are as follows:[20]

$$1,3\text{-}C_2B_7H_{13} \xrightarrow{\text{NaH}} Na_2C_2B_7H_{11}$$

$$\xrightarrow{\text{CoCl}_2} Na[Co(C_2B_7H_9)_2]$$

$$\xrightarrow{\text{Mn}_2(\text{CO})_{10}} Na[(C_2B_6H_8)Mn(CO)_3]$$

Although in all of these transition metal complexes the metal is located over a pentagonal face of the *nido* carborane anion, and metal–ligand interactions similar to those in cyclopentadienyl complexes may be envisaged for the more symmetrical species, it is not essential that the element completing the polyhedron is a transition metal. Main group metals can also complete the polyhedron. Reactions leading to icosahedral species are as follows:[21]

$$C_2B_9H_{13} + Me_2Be(OEt_2)_2 \xrightarrow{\text{benzene}} Et_2O,BeC_2B_9H_{11}$$

$$\downarrow Me_3N$$

$$Me_3N,BeC_2B_9H_{11}$$

$$Na_2C_2B_9H_{11} + MX_2 \rightarrow MC_2B_9H_{11} \quad (M = Ge, Sn, Pb)$$

The structures of the products are shown in Fig. 6–10. These may be regarded as coordination complexes of the M^{2+} cation with the anion $C_2B_9H_{11}^{2-}$ with a suitable donor species (either diethyl ether or trimethylamine) completing the coordination sphere in the case of beryllium. The lack of such a donor species in the case of the Group IV metal complexes is intelligible in that the Group IV M^{2+} cations retain a non-bonding

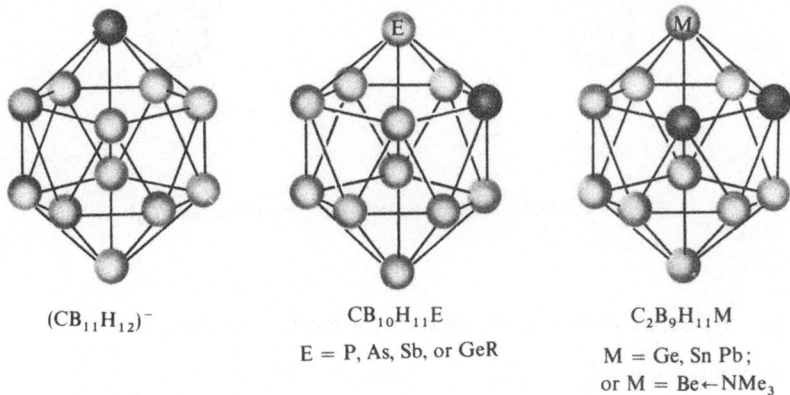

$(CB_{11}H_{12})^-$

$CB_{10}H_{11}E$

E = P, As, Sb, or GeR

$C_2B_9H_{11}M$

M = Ge, Sn Pb;
or M = Be←NMe₃

Fig. 6–10   Some isoelectronic icosahedral systems

pair of electrons which is likely to be polarized away from the dicarbollide ligand. If anything, these Group IV metal complexes might be expected to function as weak donor species, coordinating through donation of this lone pair to a suitable acceptor species. These compounds provide further examples of isoelectronic relationships, in that the species $[L \rightarrow Be]^{2+}$ or $[M]^{2+}$ (M = Ge, Sn, or Pb) occupy the icosahedral site otherwise formally occupied by $[H–B]^{2+}$ in combination with $[C_2B_9H_{11}]^{2-}$.

The C_2B_{10} icosahedron can also be regenerated by the action of the anion $C_2B_9H_{11}^{2-}$ on a suitable organoboron dihalide:

$$Na_2C_2B_9H_{11} + PhBCl_2 \rightarrow C_2B_{10}H_{11}Ph + 2NaCl$$

Such a reaction, following alkaline degradation of 1,2-dicarba-*closo*-dodecaborane, can thus be used to effect substitution of an organic group for hydrogen at the 3 (or 6) position.

An example of the incorporation of another Group III element in a carborane polyhedron is provided by the compound $MeGaC_2B_4H_6$ (1-methyl-1-galla-2,3-dicarba-*closo*-heptaborane (7), see Fig. 6–11), which is obtained by the gas-phase reaction between trimethylgallium and 2,3-dicarba-*nido*-hexaborane (8), $C_2B_4H_8$, at 215°C.[22]

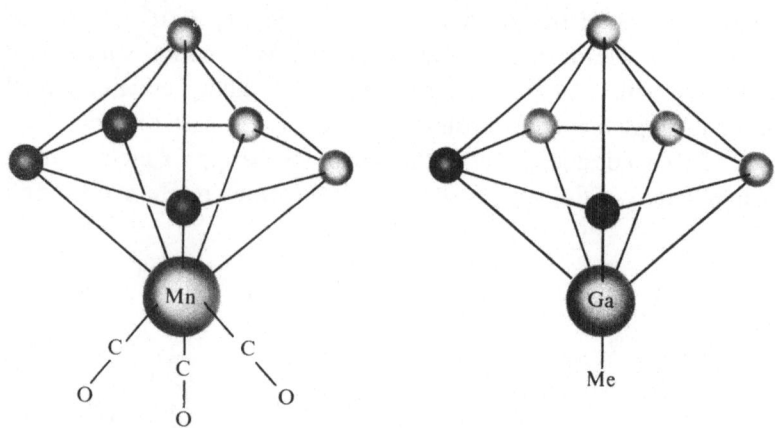

Fig. 6–11   Metallocarboranes formally derived from the *nido*-anions $C_3B_3H_6^-$ and $C_2B_4H_6^{2-}$

6–5   Species containing one cage carbon atom

The compounds so far described have all contained two cage carbon atoms. The chemistry of carboranes containing only one cage carbon atom is much less developed, but enough is known to show that the structural pattern established for the dicarba species persists for iso-electronic monocarba species, as indeed it does also for species containing more than two carbon atoms.

By analogy with borane anions $B_nH_n^{2-}$ and neutral dicarba-*closo*-boranes $C_2B_{n-2}H_n$, the isoelectronic monocarba anions $CB_{n-1}H_n^-$ may be expected to have closed cage structures based on the polyhedra listed in Table 6–1. On the other hand, *nido* structures are expected for anions $CB_{n-1}H_n^{3-}$ (cf. $C_2B_{n-2}H_n^{2-}$).

An elegant method by which a single carbon atom can be introduced into a boron cage is by exploiting the particular Lewis basicity, and unsaturation, of the cyanide ion $C\!:\!N^-$.[23] As a donor species, cyanide ion normally coordinates through the carbon atom, and with decaborane in aqueous solution it initially forms the complex $(B_{10}H_{14}CN)^-$ in which the cyanide group is attached through carbon to one of the boron atoms (probably number 6 or 9—see Fig. 3–10). This complex anion spontaneously eliminates hydrogen to form the anion $B_{10}H_{12}CN^-$, which is converted by dimethyl sulphate into the neutral species $B_{10}H_{12}CNMe_3$, a zwitterionic derivative $B_{10}H_{12}C^-\!—N^+Me_3$ of the anion $B_{10}H_{12}CH^-$: methylation of the nitrogen has reduced to a single bond the link by which it was bound to carbon, releasing the spare carbon valences to be used in cage bonding. The overall reaction sequence for preparing the anion $B_{10}H_{12}CH^-$ is as follows:

$$B_{10}H_{14} \xrightarrow{CN^-:-H_2} B_{10}H_{12}CN^- \xrightarrow{Me_2SO_4} B_{10}H_{12}CNMe_3 \xrightarrow{NaH/THF} NaCB_{10}H_{13}$$

$$\downarrow Na/THF \qquad\qquad \nearrow H_2O$$

$$Na_3CB_{10}H_{11},2THF$$

The *nido*-structure of the anion $CB_{10}H_{11}^{3-}$ was shown in Fig. 6–7. The anion $(CB_{10}H_{13})^-$ is believed to have a similar skeleton, the two extra hydrogen atoms probably bridging the boron atoms of non-adjacent edges of the pentagonal face. Oxidation of the *nido*-anion $CB_{10}H_{11}^{3-}$ converts it into the *closo* anion $CB_{10}H_{11}^-$:

$$Na_3CB_{10}H_{11},2THF + I_2 \rightarrow NaCB_{10}H_{11} + 2NaI + 2THF$$

Like its dicarba analogue $C_2B_9H_{11}^{2-}$, the anion $CB_{10}H_{11}^{3-}$ forms a wide range of transition metal complexes, for example,

$$Na_3CB_{10}H_{11} \xrightarrow{CoCl_2/THF} Co + NaCl + Na_3[Co^{III}(CB_{10}H_{11})_2]$$

similarly for Ni, Cr, Mn, and Fe.

The complexes are believed to be isostructural with their isoelectronic $C_2B_9H_{11}^{2-}$ counterparts. The monocarba anion $CB_{10}H_{11}^{3-}$ appears superior even to $C_2B_9H_{11}^{2-}$ in its ability to stabilize high metal oxidation states.

Transition metal complexes of the C-hydroxy anion $C(OH)B_{10}H_{10}^{3-}$ can be prepared directly by an ingenious modification of the route already described, in which carbon monoxide instead of cyanide ion is inserted into a borane anion:[24]

$$B_{10}H_{13}^- + M(CO)_6 \xrightarrow{u.v.} [C(OH)B_{10}H_{10}]M(CO)_4^- \quad (M = Cr, Mo, W)$$

Unlike other carboranes and metallocarboranes, these C-hydroxy species undergo base degradation with loss of a carbon atom, *not* a boron atom, from the icosahedral framework to generate *nido* metalloboranes $B_{10}H_{12}M(CO)_4^{2-}$.

The triply-charged *nido* anions $CB_{10}H_{11}^{3-}$ are convenient species from which to prepare carba*phospha*boranes and related substances $CHEB_{10}H_{10}$ (E = P, As, Sb, or MeGe):[25]

$$Na_3CB_{10}H_{11},2THF \xrightarrow{ECl_3} 1,2\text{-}CHEB_{10}H_{10} \xrightarrow{450-500°} 1,7\text{-}CHEB_{10}H_{10}$$

These compounds may be regarded as derived from the C_2B_{10} carborane by replacing one CH group by a phosphorus, arsenic, or antimony atom or by an alkylgermanium group RGe (see Fig. 6–10). They are very similar in their physical and chemical properties to the isoelectronic dicarba-*closo*-dodecaboranes, undergoing comparable isomerization and degradation reactions:

$$1,2\text{-CHEB}_{10}\text{H}_{10} \xrightarrow{\text{piperidine}} 1,2\text{-CHEB}_9\text{H}_{10}^- \qquad E = P, As, Sb$$

$$1,7\text{-CHEB}_{10}\text{H}_{10} \xrightarrow{\text{piperidine}} 1,7\text{-CHEB}_9\text{H}_{10}^- \qquad E = P \text{ or } As.$$

The *nido* anions $\text{CHEB}_9\text{H}_{10}^-$ that result from these base degradations form the usual range of transition metal complexes in which the metal completes the icosahedron; derivatives in which the metal is coordinated to the lone pair of the Group V element are also known.

Closo-carboranes with CB_9 and CB_{11} frameworks result from thermal disproportionation of the CB_{10} species $\text{CsCB}_{10}\text{H}_{13}$:[26]

$$2\text{CsCB}_{10}\text{H}_{13} \xrightarrow{300-320°} \text{CsCB}_9\text{H}_{10} + \text{CsCB}_{11}\text{H}_{12} + 2\text{H}_2$$

The [11]B n.m.r. spectra of the products show that the anions CHB_9H_9^- and $\text{CHB}_{11}\text{H}_{11}^-$ probably have the expected *closo* geometry:

CsCHB_9H_9: 3 doublets, of area ratio $1:4:4$, as expected for a bicapped Archimedean antiprism with one apical carbon atom

$\text{CsCHB}_{11}\text{H}_{11}$: 3 doublets, of area ratio $1:5:5$, as required for an icosahedral monocarborane

Lower carboranes containing only one cage carbon atom are usually made from a pentaborane or smaller boron hydride,[8,11,27] for example,

$$1\text{-MeB}_5\text{H}_8 \xrightarrow[\text{or strong heat}]{\text{electric discharge}} \text{CB}_5\text{H}_7$$

$$\text{B}_5\text{H}_9 + \text{LiC}\text{:Me} \rightarrow \text{EtCB}_5\text{H}_8 \text{ (a C-alkylated derivative of } \text{CB}_5\text{H}_9,$$
$$\text{see Fig. 6-3)}$$

$$\text{EtBF}_2 + \text{Li} \xrightarrow{\text{THF; 0°}} \text{various products, including alkyl derivatives of}$$
$$\text{CB}_5\text{H}_9$$

$$\text{C}_2\text{B}_6\text{H}_8 \xrightarrow[\text{2. HCl gas}]{\text{1. Me}_4\text{NBH}_4 \text{ in diglyme}} \text{CB}_5\text{H}_9 \text{ and alkyl derivatives thereof}$$

This last reaction is a rare example of the removal of a carbon atom from a closed carborane polyhedron. Note that these reactions all lead to CB_5 species. The pentagonal pyramidal structure of the compound CB_5H_9 is shown in Fig. 6-3. The compound with two fewer hydrogen atoms, CB_5H_7, which may be thought of as formally derived from the *closo* anion CB_5H_6^-, has an octahedral skeleton with one bridging hydrogen atom, as shown by the following spectroscopic evidence:

Infrared spectrum: $v(\text{C-H})$ 2960; $v(\text{B-H})$ 2640; $v(\text{BHB})$ 2167 cm^{-1}.

[11]B n.m.r. spectrum: three doublets, of area ratio $2:2:1$, with coupling constants appropriate for terminal B-H groups.

6-6   Carboranes with more than two cage carbon atoms

Few carboranes are yet known with more than two cage carbon atoms. Two such species, $\text{C}_3\text{B}_3\text{H}_7$ and $\text{C}_4\text{B}_2\text{H}_6$, are members of the isoelectronic series based on hexaborane(10), B_6H_{10}, and are believed to have pentagonal-based pyramidal structures (see Fig. 6–3).

The tricarba compound $C_3B_3H_7$ and a number of its C-methyl derivatives are formed in exceedingly low yield in the vapour phase reaction between tetraborane(10) and acetylene at 25°.[28] Absorptions near 1900 and 1500 cm^{-1} in the infrared spectrum are indicative of bridging hydrogen atoms, and in the ^{11}B n.m.r. spectrum, a doublet at high field is assignable to the apical boron, while a doublet at lower field, of twice the intensity and with doublet fine structure, is appropriate for two basal BH groups linked by a bridging hydrogen.

Removal (as a proton) of the bridging hydrogen of tricarba-*nido*-hexaborane(7), $C_3B_3H_7$, gives the anion $C_3B_3H_6^-$, the pentagonal face of which can coordinate to transition metals (see Fig. 6–11):

$$\tfrac{1}{2}Mn_2(CO)_{10} + MeC_3B_3H_6 \xrightarrow{175-200°} (MeC_3B_3H_5)Mn(CO)_5 + 2CO + \tfrac{1}{2}H_2$$

Several alkyl derivatives of tetracarba-*nido*-hexaborane, $C_4B_2H_6$ (Fig. 6–3), have been prepared starting from alkynylboranes,[11] for example,

$$Na[Me_3BC\!:\!CMe] \xrightarrow{Me_2BCl}$$

$$\xrightarrow{-Me_3B} Me_6C_4B_2$$

The parent compound is obtained in low yield by passing a stream of 1,2-tetramethylenediborane(6) at low pressure through a tube at 550°:[29]

$\xrightarrow{550°}$ some $C_4B_2H_6$ (and other carboranes)

It is particularly interesting that these compounds have the pentagonal pyramidal structure (Fig. 6–3, shown in plan as [1] above) rather than the six-membered ring structure [2], which might have seemed likely to result from the former reaction, but which would however be more appropriate for a species with two more bonding electrons, $R_6C_4B_2^{2-}$, being then isoelectronic with benzene. The ^{11}B and ^1H n.m.r. spectra of these compounds, however, rule out structure [2], as they indicate the presence of two types of boron atom and two types of skeletal carbon atom in the proportions appropriate for [1].

References

1. Bobinski, J., *J. Chem. Educ.*, 1964, **41**, 500.
2. Brotherton, R. J., and H. Steinberg (eds.), *Progress in boron chemistry*; Vol. II, Pergamon Press, Oxford, 1970.
3. Williams, R. E., pp. 37–118 of ref. 2.
4. Heying, T. L., pp. 119–139 of ref. 2.
5. Lipscomb, W. N., *Science*, 1966, **153**, 373.
6. Muetterties, E. L., and W. H. Knoth, *Polyhedral Boranes*, Marcel Dekker, New York, 1968.
7. Bregadze, V. I., and O. Yu. Okhlobystin, *Organometal Chem. Rev.*, 1969, **4A**, 345.
8. Onak, T. P., *Advances Organometal Chem.*, 1965, **3**, 263.
9. Issleib, K., R. Linder, and A. Tzschach, *Z. Chem.*, 1966, **6**, 1.
10. Hawthorne, M. F., *Endeavour*, 1966, **25**, 146.
11. Köster, R., and M. A. Grassberger, *Angew. Chem. Internat. Edn.*, 1967, **6**, 218.
12. For annual surveys of developments see *Annual Reports Chem. Soc. (London)* and *Organometal Chem. Rev., Section B*.
13. Hart, H. V., and W. N. Lipscomb, *J. Amer. Chem. Soc.*, 1969, **91**, 771.
14. Smith, H. D., *Inorg. Chem.*, 1969, **8**, 676.
15. Tebbe, F. N., P. M. Garrett, and M. F. Hawthorne, *J. Amer. Chem. Soc.*, 1968, **90**, 869; Garrett, P. M., G. S. Ditta, and M. F. Hawthorne, *Inorg. Chem.*, 1970, **9**, 1947.
16. Ditter, J. F., E. B. Klusmann, J. D. Oakes, and R. E. Williams, *Inorg. Chem.*, 1970, **9**, 889.
17. Hawthorne, M. F., D. C. Young, J. D. Andrews, D. V. Howe, R. L. Pilling, A. D. Pitts, M. Reintjes, L. F. Warren, and P. A. Wegner, *J. Amer. Chem. Soc.*, 1968, **90**, 879; Warren, L. F., and M. F. Hawthorne, *ibid.*, 1970, **92**, 1157.
18. Wing, R. M., *J. Amer. Chem. Soc.*, 1967, **89**, 5599; 1968, **90**, 4828; 1970, **92**, 1187.
19. St. Clair, D., A. Zalkin, and D. H. Templeton, *Inorg. Chem.*, 1969, **8**, 2080.
20. George, A. D., and M. F. Hawthorne, *Inorg. Chem.*, 1969, **8**, 1801; George, T. A., and M. F. Hawthorne, *J. Amer. Chem. Soc.*, 1969, **91**, 5475.
21. Rudolph, R. W., R. L. Voorkees, and R. E. Cochoy, *J. Amer. Chem. Soc.*, 1970, **92**, 3351.
22. Grimes, R. N., and W. J. Rademaker, *J. Amer. Chem. Soc.*, 1969, **91**, 6498.
23. Scholer, F. R., and L. J. Todd, *J. Organometal Chem.*, 1968, **14**, 261.
24. Wegner, P. A., L. J. Guggenberger, and E. L. Muetterties, *J. Amer. Chem. Soc.*, 1970, **92**, 3473.
25. Todd, L. J., A. R. Burke, A. R. Garber, H. T. Silverstein, and B. N. Storhoff, *Inorg. Chem.*, 1970, **9**, 2175.
26. Knoth, W. H., *J. Amer. Chem. Soc.*, 1967, **89**, 1274.
27. Onak, T., P. Mattschei, and E. Groszek, *J. Amer. Chem. Soc.*, 1969, **91**, 1990; Dunks, G. B., and M. F. Hawthorne, *Inorg. Chem.*, 1969, **8**, 2667.
28. Franz, D. A., J. W. Howard, and R. N. Grimes, *J. Amer. Chem. Soc.*, 1969, **91**, 4010, 6499.

29. Onak, T. P., and G. T. F. Wong, *J. Amer. Chem. Soc.*, 1970, **72**, 5226.
30. Williams, R. E., *Inorg. Chem.*, 1971, **10**, 210.
31. Wade, K., *Chem. Comm.*, 1971, 792.
32. Grimes, R. N., *Carboranes*, Academic Press, New York, 1971.
33. Snaith, R. and K. Wade, Carboranes and metallocarboranes, in Lappert, M. F., ed. *Internat. Rev. Sci., Inorg. Chem.*, Vol. 1. M.T.P. and Butterworth, London, 1972.

Problems

6–1 What isomeric forms of dicarba-*closo*-heptaborane(7) can in principle exist? How many types of carbon and boron atom does each contain? What absorptions might be expected in their ^{11}B and ^{1}H n.m.r. spectra?

6–2 What isomeric forms of dimethyl-1,6-dicarba-*closo*-hexaborane(6) can in principle exist? Explain to what extent these should be distinguishable from each other by means of their ^{11}B and ^{1}H n.m.r. spectra.

Carry out the same exercise for the isomeric forms of dimethyl-1,2-dicarba-*closo*-hexaborane(6).

6–3 Discuss the types of cage azaborane that might in principle be able to exist if one or two nitrogen atoms could be incorporated in cages otherwise composed exclusively of boron atoms.

6–4 Suggest likely structures for the following species:

$$CB_4H_5^- \; ; \; CB_4H_5^{3-} \; ; \; C_4B_2H_6^{2-} \; ; \; CB_8H_9^- \; ; \; MeAlC_2B_4H_6 \; ; \; CBH_6^- \; ;$$

$$CB_8H_9^{3-} \; ; \; [(C_2B_4H_6)_2Fe(II)]^{2-}$$

6–5 Explain what absorptions may be expected in the ^{11}B and ^{1}H n.m.r. spectra of (a) tetracarba-*nido*-hexaborane(6) and (b) hexamethyl-tetracarba-*nido*-hexaborane(6). What features of these spectra allow these compounds to be distinguished from hypothetical isomers with six-membered ring structures [2]?

6–6 An equimolar mixture of diborane and acetylene diluted with hydrogen and subjected to electric discharge at 25° gave a mixture of volatile carboranes and methyl derivatives of carboranes. One of these had a molecular weight of 89·5 and contained 10% by weight of hydrogen. A peak at 2570 cm^{-1} was the only absorption in its infrared spectrum between 1500 and 2800 cm^{-1}. Its ^{11}B n.m.r. spectrum consisted of three absorptions of equal intensity; a singlet at low field, a doublet next, and another doublet at highest field. In its ^{1}H n.m.r. spectrum, the major features were three singlets of relative intensities 1:3:3, at fields appropriate for carbon-attached hydrogen atoms.

Suggest a possible structure or structures for this compound, explaining the steps in your reasoning and the significance of the various pieces of spectroscopic information given.

7 Other examples of electron deficient bonding

7–1 Introduction

The criterion of electron deficiency that has been used in this book is that a molecule or ion possesses too few bonding electrons to allow a pair to be allocated to every pair of atoms close enough to be regarded as covalently bonded. The types of compound covered by this definition have included at one extreme substances like diborane, in which the electron deficiency is effectively localized in one section of the molecule, a section which accordingly serves as a point of attack by electron-rich species. At the other extreme are substances like borane anions $B_nH_n^{2-}$ and carboranes $C_2B_nH_{n+2}$, in which the electron deficiency is spread over the whole of the molecular skeleton, which therefore presents no localized reactive site for attack by nucleophiles.

This final chapter contains a brief survey of some other systems which formally are included by the same definition of electron deficiency. For example, as outlined in Section 7–2 below, certain carbonium ions adopt structures in which three-centre, two-electron bonds (CCC, CHC, or HCH bonds) are helpful in a discussion of their bonding, in which respect these ions have a close similarity to the lower boranes. On the other hand, the polyhedral atom clusters that characterize the structures of borane anions, carboranes, and lithium alkyls have counterparts in a number of transition metal cluster compounds, some examples of which are described in Section 7–3. The interaction between *nido*- carborane anions $C_2B_{n-2}H_n^{2-}$, etc., and transition metals has already been noted to resemble metal–cyclopentadienyl interactions, a point which is elaborated briefly in Section 7–4. Also included in this chapter is a discussion (Section 7–5) of some systems which may well be held together by low-order sigma bonds despite the apparent availability of enough valence electrons to allow a pair for each two-centre link.

7–2 Electron deficient bridged carbonium ions

Carbonium ions, the singly charged organic cations R^+ that may result from the separation of negatively charged atoms or groups X^- from neutral molecules RX, or from the protonation of organic substances, are important intermediates in many organic reactions. A knowledge of their structures is clearly vital to an understanding of these reactions. Few carbonium ions, however, are so unreactive as to be isolable as salts which can be characterized and studied at normal temperatures. Information on their structures usually has to be inferred from the products of their reactions with nucleophiles, or derived from low temperature spectroscopic studies on solutions in strongly acid media. As both reaction products and spectra can often be rationalized in terms of more than one structure, it is not surprising that there has been considerable discussion concerning the identity of a number of key species, revolving particularly around the question of whether they are best represented by a single structure involving three-centre bonds, or more realistically taken to be a rapidly equilibrating mixture of two or more distinct species for which two-centre bonding descriptions are adequate.

One system for which two-centre bonding descriptions are obviously inadequate is the methanonium ion CH_5^+ and its higher homologues $[C_2H_{2n+3}]^+$ which are now known to be produced, at least initially, when alkanes are dissolved in strongly acid media. The parent ion, CH_5^+, is well known in the gas phase from mass spectroscopic studies of methane,[1] which under electron impact fragments to such ions as CH_x^+ $(x = 0 \rightarrow 4)$ and H_y^+ $(y = 1$ or $2)$ as primary products. These in turn may react with other methane molecules to give various secondary products, including the methanonium ion:

$$CH_4^+ + CH_4 \xrightarrow{-CH_3} CH_5^+ \rightleftharpoons CH_3^+ + H_2$$

With six atoms held together by means of four pairs of electrons, the ion CH_5^+ clearly falls within our definition of electron deficiency. It probably has a trigonal bipyramidal arrangement of the five hydrogen atoms about the central carbon (Fig. 7–1a) which in a simple bonding picture could be taken as sp^2 hybridized, with one sp^2 hybrid orbital available for a two-centre bond to each equatorial hydrogen, and a p orbital for a three-centre HCH bond to the axial hydrogen atoms. An alternative possible structure

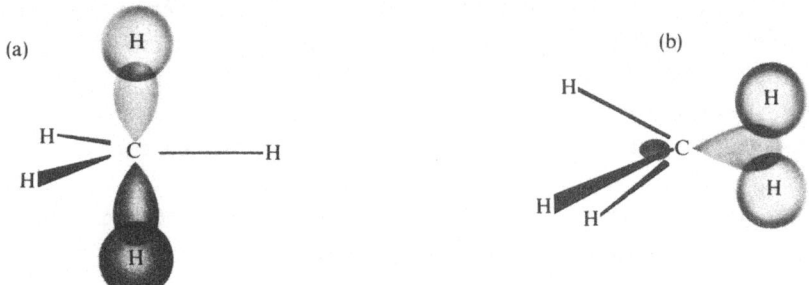

(a)   (b)

Fig. 7–1   Alternative possible structures for the ion CH_5^+:
(a) the vacant p orbital of a planar CH_3^+ ion interacts with the s orbitals of two axial hydrogen atoms;
(b) the vacant sp^3 hybrid orbital of a pyramidal CH_3^+ ion interacts with an elongated hydrogen molecule.

is that shown in Fig. 7–1b, in which a pyramidal CH_3^+ ion, which may be regarded as held together by three two-centre CH bonds using three sp^3 hybrid carbon orbitals, is so orientated with respect to an elongated hydrogen molecule as to allow the vacant carbon sp^3 hybrid orbital to interact with the H–H bonding orbital, forming a central HCH three-centre bond.

The stability of this carbonium ion in the gas phase is underlined by the enthalpy of its dissociation $CH_5^+ \rightarrow CH_3^+ + H_2$, for which ΔH has been estimated[2] to be about $+150$ kJ/mole, that is the dissociation is *endo-thermic*. An even greater enthalpy change, ΔH, of about $+500$ kJ/mole has been estimated for the decomposition into methane and a proton, $CH_5^+ \rightarrow CH_4 + H^+$. These enthalpy changes, however, relate to gas phase reactions. In solution in polar solvents, the solvation energy of the ions CH_3^+ or H^+ will considerably exceed that of the ion CH_5^+. Nevertheless, it is not surprising that methane behaves as a base in strongly acid media, in which such ions as CH_5^+ appear to have a real, if short-lived, existence.[2] The high solubility of methane in sulphuric acid was noticed as long ago as 1899,[3] though it is only recently that its behaviour in strong acids has been studied in any detail.[2] In a 1:1 mixture of fluorosulphuric acid and antimony pentafluoride (G. A. Olah's 'magic acid') at 140°C methane is apparently protonated to the methanonium ion CH_5^+, which subsequently decomposes by a succession of hydrogen elimination reactions and reactions with further methane molecules, affording the stable t-butyl cation $(Bu^t)^+$ as the major product:

$$CH_4 \xrightarrow{\ H^+\ } CH_5^+ \xrightarrow{\ -H_2\ } CH_3^+ \xrightarrow{\ CH_4\ } C_2H_7^+ \xrightarrow{\ -H_2\ } C_2H_5^+ \longrightarrow \text{etc.} \longrightarrow C_4H_9^+$$

For all the intermediates $[C_nH_{2n+3}]^+$ believed to be involved in these reactions, structures analogous to that in Fig. 7–1, with one five-coordinate carbon atom, are proposed.

It is the carbonium ions of general formula $[C_nH_{2n+1}]^+$, however, that are much more often involved in organic chemical reactions, and the structural problems they pose can be illustrated by considering the ethyl cation, $C_2H_5^+$, as a simple example. This behaves in a manner consistent with relatively ready exchange of hydrogen atoms between one carbon atom and the other (as can be shown by preparing the ion from partially deuterated ethyl derivatives, for example, $^2H_3CC^1H_2X$ or $^1H_3CC^2H_2X$), a process for which a hydrogen bridged (protonated ethylene) intermediate may be envisaged, with a CHC bridge bond like a BHB bridge bond of diborane:

$$H_3\overset{+}{C}-CH_2 \rightleftharpoons H_2C\overset{\overset{\overset{+}{H}}{\frown}}{\quad\quad} CH_2 \rightleftharpoons H_2\overset{+}{C}-CH_3$$

This system is simple enough for a realistic assessment of the relative energies of these species to be calculated for various plausible values of the interatomic distances and bond angles. When this is done,[4] it appears that the protonated ethylene structure has an energy some 14 kJ/mole *higher* than that of the ethyl cation, and indeed higher than the energies of species with structures intermediate between these two. That is, the protonated ethylene structure, rather than being an *intermediate* in the hydrogen exchange, is effectively the *transition state*, as it represents an energy maximum on the reaction coordinate (see Fig. 7–2). It therefore appears that in this case, the carbonium ion is better described as an equilibrating pair of ethyl cations, that is, in terms of structures which

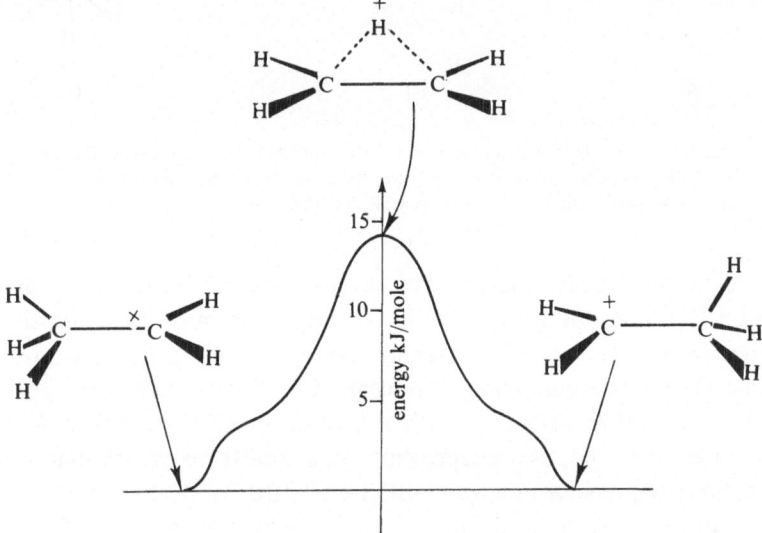

Fig. 7–2 The calculated relative energies of the ethyl cation $[H_3CCH_2]^+$ and protonated ethylene $[H_2CHCH_2]^+$ structures for the ion $C_2H_5^+$

can be adequately represented in terms of localized two-centre electron pair bonds, rather than in terms of a structure containing a three-centre CHC bond. The energy difference between the two types of structure is small, however, and structures containing three-centre CHC bridge bonds may still be possible for higher carbonium ions.

Calculations on the fluoroethyl cation $C_2H_4F^+$ show that in this system, the protonated fluoroethylene structure $(FHCHCH_2)^+$ is probably some 50 kJ mole *more* stable than the 2-fluoroethyl cation $(FH_2CCH_2)^+$, though about 120 kJ/mole *less* stable than the 1-fluoroethyl cation $(H_3CCHF)^+$, which is accordingly the preferred structure.[5]

Carbonium ions containing at least three carbon atoms offer scope for structures with three-centre CCC bonds, whether as low energy transition states whereby one two-centre bonded structure can change into another,

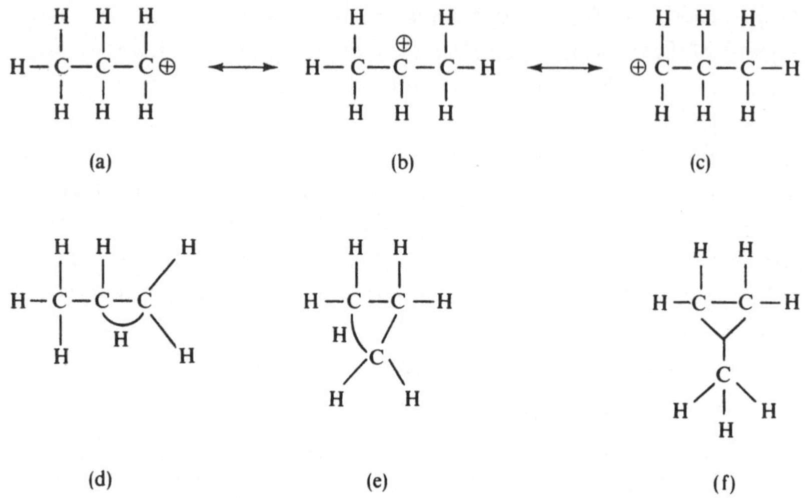

Fig. 7–3 Possible structures for a propyl cation $C_3H_7^+$. In structures (a)–(c), the bonding may be described in terms of localized two-centre electron pair bonds. Structures (d)–(f) involve three-centre CHC or CCC bonds.

or as the preferred structure. The diverse possibilities for the propyl cation $C_3H_7^+$ are illustrated in Fig. 7–3. Apart from the three possible structures containing only two-centre bonds (Fig. 7–3a–c) linear (d) or cyclic (e) structures containing three-centre CHC hydrogen bridges appear feasible, as well as a structure with a three-centre CCC bond (Fig. 7–3f). The three-centre bonds are represented in a similar manner to that used in earlier chapters for analogous BHB and BBB bonds. Even if the last three three-centre bonded ('non-classical') structures are not of lower energy than the 'classical' structures with only two-centre bonds, they clearly afford low energy intermediates or transition states through which scrambling reactions can occur.

The rapidity with which scrambling can occur has been revealed by a detailed mass spectroscopic study of the butyl cations $C_4H_9^+$ generated from partially deuterated n-, iso-, s-, and t-butyl compounds.[6] After only a few microseconds, the various cations have isomerized almost completely to the same (unknown) structure or mixture of structures before losing ethylene or methane with a rate constant in the region 10^5–10^6 s^{-1}.

Useful information on the structures of certain relatively stable carbonium ions in solution has been obtained from ^{1}H and ^{13}C spectroscopic studies using the fluorosulphuric acid–antimony pentafluoride solvent system with sulphuryl chloride fluoride as a diluent at low temperatures.[7] Primary and secondary carbonium ions, RCH_2^+ and R_2CH^+, isomerize or otherwise decompose too rapidly to be studied in this solvent system—both ethane and propane for example give mixtures of t-butyl and -hexyl cations—but tertiary carbonium ions are stable, and crystalline derivatives of the ions Me_3C^+, $Me_2\overset{+}{C}Et$ and $Me_2HC\overset{+}{C}Me_2$ have been prepared. In solution, the t-butyl cation gives the expected singlet in the ^1H n.m.r. spectrum. The hexyl cation $Me_2CH\overset{+}{C}Me_2$ gives the spectrum shown in Fig. 7–4, which shows the four methyl groups to be equivalent,

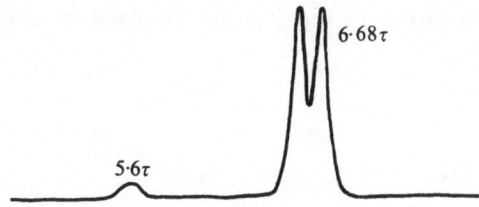

6·68τ

5·6τ

Fig. 7–4   The ^1H n.m.r. spectrum of the ion $Me_2HCCMe_2^+$

coupled equally to the methine proton and so giving a doublet signal. This spectrum in form corresponds to that expected for a protonated tetramethylethylene structure, $(Me_2CHCMe_2)^+$, but as the chemical shift of the methine proton is appropriate for hydrogen attached to a carbon atom α to a carbonium carbon (that is, it is appropriate for the system $Me_2CH-\overset{+}{C}\!\!<$), the spectrum is interpreted as indicating that the 1,1,2,2-tetramethylethyl cation structures $Me_2HC\overset{+}{C}Me_2 \rightleftharpoons Me_2\overset{+}{C}CHMe_2$ are adopted in this case, equilibrating rapidly enough for the four methyl groups to appear equivalent.

The pentamethylethyl cation $Me_5C_2^+$, formed by dissolving the heptane Me_3CCHMe_2 in the mixed acid solvent at $-60°$, gives only one singlet in the ^1H n.m.r. spectrum, showing that all five methyl groups are magnetically equivalent. This rules out a static, methyl-bridged structure $(Me_2CMeCMe_2)^+$, for which two signals, of relative areas 4:1, would be expected. The chemical shift is moreover that appropriate for terminal, rather than partially bridging, methyl hydrogens. The barrier to exchange, $Me_3C\overset{+}{C}Me_2 \rightleftharpoons Me_2\overset{+}{C}CMe_3$, is estimated to be less than about 10 kJ/mole.

Bridged structures in which a halogen atom (Cl, Br, or I, but not F) is bridging are probably adopted by the ions $Me_2CXCMe_2^+$, which give only singlet absorptions even at $-95°$:[8]

$$Me_2\overset{XX}{C}CMe_2 \xrightarrow{SbF_5/SO_2} [Me_2C\overset{\displaystyle X}{\triangle}CMe_2]^+$$

In these systems, however, the halogen atom can supply three electrons towards the bridge link, which is therefore not electron deficient.

Carbonium ions derived from cyclic starting materials may adopt cage structures during isomerization reactions. For example, whether one starts from cyclopropylmethyl compounds $CH_2CH_2CHCH_2X$ or from cyclobutyl compounds $CH_2CH_2CH_2CHX$, loss of halide or toluene sulphonate ion X^- apparently gives the same carbonium ion(s), since both afford similar proportions of cyclopropylmethyl and cyclobutyl derivatives among the solvolysis products. It appears likely that pyramidal species of the type shown in Fig. 7–5 are involved in these reactions.[9]

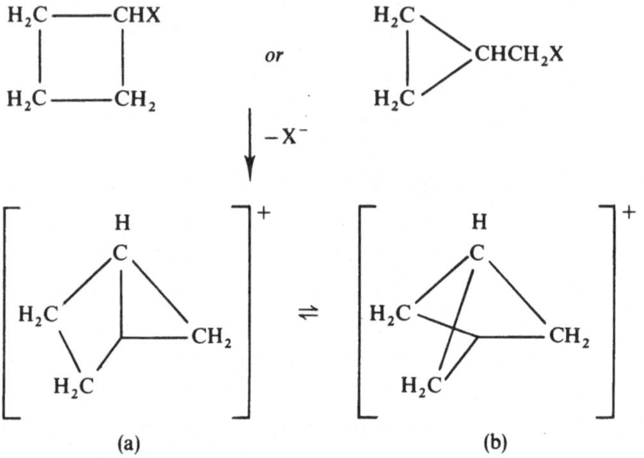

Fig. 7–5   Plausible canonical forms for the pyramidal carbonium ion believed to feature in certain reactions of cyclobutyl or cyclopropylmethyl derivatives

The cyclic systems that have commanded most attention are those containing the bicyclo[2,2,1]heptane bridged ring system, the literature on which is now vast.[10] It was early recognized that the stable 2-norbornyl cation, which is accessible by the various routes outlined in Fig. 7–6, reacted in a manner consistent with a three-centre bonded structure or with a rapid equilibration between various two-centre bonded structures. That it is probably best described as a corner protonated nortricyclene (Fig. 7–7a–c), a protonated cyclopropane (cf. Fig. 7–3f) containing a five-coordinate carbon atom, is indicated by its Raman, ^1H and ^{13}C n.m.r. spectra, which were described by G. A. Olah in his hundredth research paper[11] on carbonium ions. The ^1H n.m.r. spectra at $-113°$ and $-154°C$ are shown in Fig. 7–7. Although the four hydrogen atoms attached to the three carbon atoms of the cyclopropane ring section of the molecule are indistinguishable in the ^1H n.m.r. spectrum at $-113°C$, at lower temperatures they give rise to two peaks of equal area, with chemical shifts appropriate for the two methine (CH) groups and one methylene (CH_2) group of a corner protonated nortricyclene. The activation energy for equilibration of these hydrogens by hydride shift around

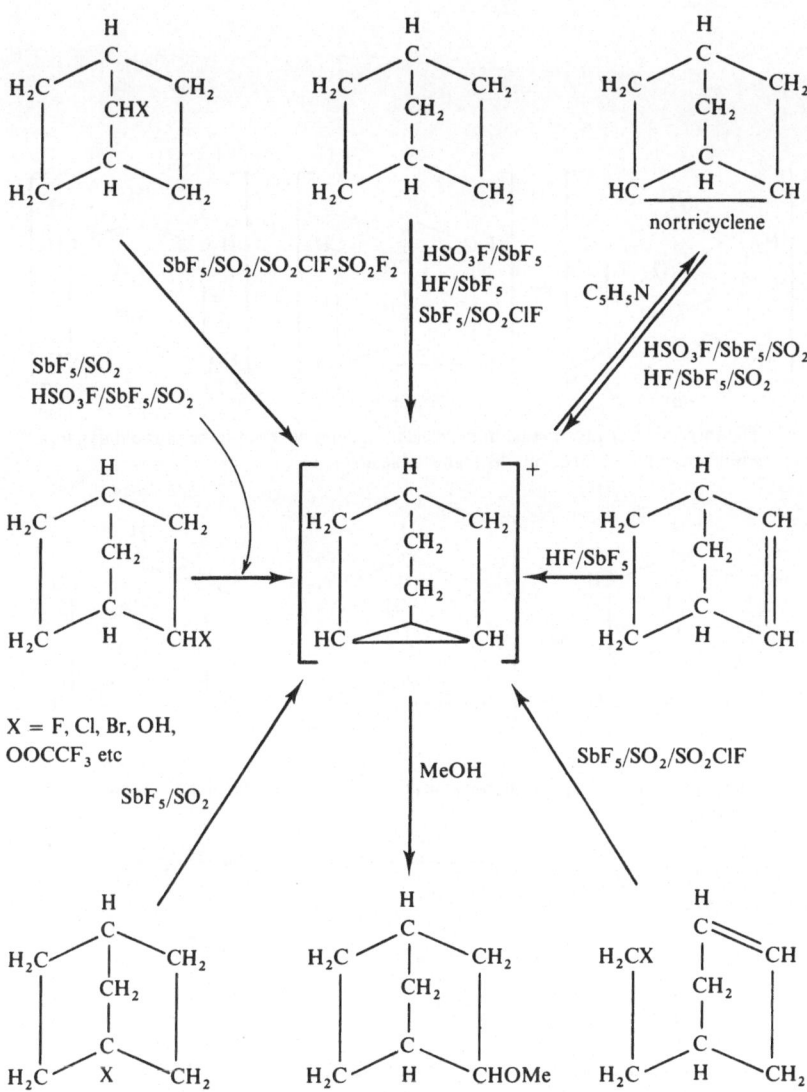

Fig. 7–6 Various methods by which the stable norbornyl cation may be prepared.

the ring is about 25 kJ/mole. The signal due to the methylene protons adjacent to the apical methine group also separates at $-154°$ into two components, though of relative intensities 2:1, as one methylene group then differs from the other two. If the ion had a static 'classical' structure as shown in Fig. 7–7d or e, the three types of ring proton would be expected to give three resonances of relative areas 2:1:1.

There remains the question of whether the ion at $-154°$ should be regarded as a rapidly equilibrating pair of 'classical' ions (Fig. 7–7d and e) or as the 'non-classical' structure Fig. 7–7b, which represents the inter-

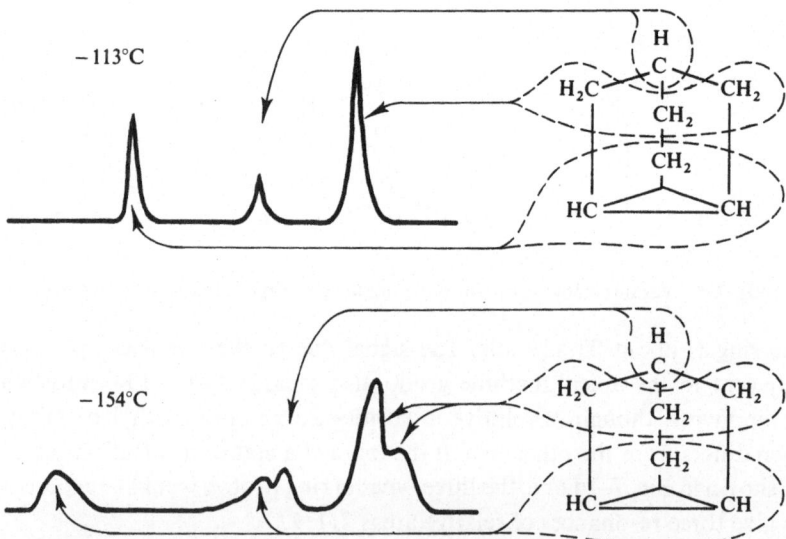

The three 'corner protonated nortricyclene' species believed to be in rapid dynamic equilibrium at $-113°C$, but distinguishable at $-154°C$.

'Classical' [(d) and (e)] and 'non-classical' (b) representations of the bonding in the norbornyl cation.

Effect of temperature on the ^1H n.m.r. spectrum of the norbornyl cation.

Fig. 7–7 Structures in terms of which the ^1H n.m.r. spectrum of the norbornyl cation may be interpreted.

mediate in such an equilibration. These alternative descriptions do not differ meaningfully unless the positions of the atomic nuclei of forms (b), (d), and (e) differ significantly—they are otherwise different resonance canonical forms of the same species. As the Raman spectrum is consistent with a symmetry appropriate for structure (b) rather than (d) or (e), the three-centre bonded structure is preferred.[11]

7–3   Transition metal cluster compounds

One of the major growth areas within the general field of transition metal chemistry since the mid-fifties has been the chemistry of compounds containing metal–metal bonds. These range from binuclear species held together by essentially localized two-centre M–M bonds, to compounds containing polyhedra of metal atoms, like the polyhedra of boron atoms that form the skeletons of boron cage species; for these, localized bonding descriptions tend to be less satisfactory than delocalized molecular orbital treatments. Among metal–metal bonded compounds, it has proved convenient to treat as a class, referred to as metal atom cluster compounds, those which contain groups of three or more metal atoms each of which is bound to all or most of the others. Even binuclear species are frequently included as special, though atypical, cases. The term 'cluster' is used in preference to 'cage' because the latter, which has acquired general usage in connection with borane anions $B_nH_n^{2-}$ and carboranes $C_2B_{n-2}H_n$, would not be appropriate for the triangular and certain other open networks of atoms that it is useful to include within the same general classification, and moreover appears to imply that another atom might be held within the cage, whereas only in very few cases is there actually room for such a possibility.

Among transition metal cluster compounds, species based on triangles, tetrahedra, or octahedra of three, four, or six metal atoms respectively are particularly common, and many other structural types are known. Few such compounds are electron deficient in the sense of possessing too few valence electrons to allow two per two-centre metal–metal link, though for the remainder descriptions of the bonding in terms of localized two-centre bonds are rarely satisfactory. Treatments using partially or completely delocalized molecular orbitals are needed to interpret such features as their metal–metal bond distances, their magnetic properties, and their electronic spectra. Our concern here is to consider a few typical examples to illustrate their relationship to cage species referred to in earlier chapters. Detailed treatments will be found in refs 12–14.

Metal atom clusters are found principally among two classes of transition metal compound—lower metal halides and metal carbonyls. They are common in the lower halide chemistry of niobium, tantalum, molybdenum, and tungsten, and in the carbonyl chemistry of elements further to the right in the transition series, such as those of the iron, cobalt, and nickel subgroups. Among lower halide clusters, two important structural types are those containing M_6X_8 and M_6X_{12} units (see Fig. 7–8).

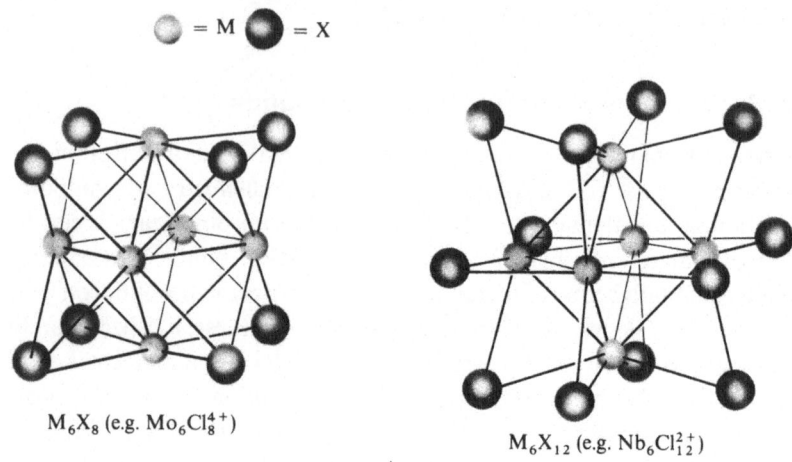

$\bigcirc$ = M $\bullet$ = X

M_6X_8 (e.g. $Mo_6Cl_8^{4+}$)

M_6X_{12} (e.g. $Nb_6Cl_{12}^{2+}$)

Fig. 7–8   Two types of octahedral metal cluster species common among the lower halides of niobium, tantalum, molybdenum, and tungsten.

The former (M_6X_8) contain an octahedral array of six metal atoms, with a triply bridging halogen atom X located over the centre of each of the eight octahedral faces. The latter (M_6X_{12}) also have an octahedral core of metal atoms, with the twelve doubly bridging halogen atoms located over the centres of the twelve octahedral edges.

Typical examples of such species are the ionic clusters $Mo_6Cl_8^{4+}$ and $Nb_6Cl_{12}^{2+}$. Their bonding can be described by approaches similar to those used for the octahedral borane anion $B_6H_6^{2-}$, although molecular orbital treatments are rendered more complicated than for the borane anion because the halogen atoms are not located on the same symmetry axes as the metal atoms, and because there are nine orbitals on each metal to consider (one s, three p, and five d orbitals) instead of only four for boron clusters. The problem can be simplified by taking account of the effectively square planar arrangement of halogen atoms about each metal atom in both types of structure (slightly distorted in the case of M_6X_{12}, see Fig. 7–8). If one d, the s, and two p orbitals (effectively in dsp^2 hybridization) are used in MX_4 metal–halogen bonds, there remain one p and four d orbitals on each metal atom. One orbital (effectively a pd hybrid) points away from the cluster along the four-fold axis, and is available for bonding to another ligand along this axis (see below). The remaining four orbitals point towards the other metals or towards the centre of the cluster, and are available for metal–metal bonding. The precise energy sequence of the molecular orbitals that result from their interactions is difficult to estimate because of uncertainties about values of the radial wave function of the d orbitals, but it has been concluded that for the species M_6X_8, there are twelve cluster metal–metal bonding orbitals (of symmetries A_{1g}, T_{1u}, T_{2g}, E_g, and T_{2u} in order of increasing energy), whereas for the species M_6X_{12}, there are eight cluster metal–metal bonding orbitals (of symmetries A_{1g}, A_{2u}, T_{1u}, and T_{2g} again in order of increasing energy).[15] The difference between these numbers and types of bonding orbitals arises because, though in both cases there is an octahedral arrangement of the metal atoms, there is a difference between the orientations of the square-planar MX_4 units with respect to the four-fold axes of the octahedra (see Fig. 7–8).

That these orbitals are appropriate in number for the number of bonding electrons available in the ions $Mo_6Cl_8^{4+}$ and $Nb_6Cl_{12}^{2+}$ can be seen as follows. The number of electrons available for metal–metal cluster bonding is taken as the number of electrons left on the metal cluster if it is written in the form M_6^{n+}, all the halogen atoms having been removed as halide ions, for example, Mo_6^{12+} in the case of $Mo_6Cl_8^{4+}$, and Nb_6^{14+} in the case of $Nb_6Cl_{12}^{2+}$. This is not to say that the metal–halogen bonding is ionic in character. This is simply a device to exclude from the electron-count the two or three pairs of electrons by which each halogen performs a doubly or triply bridging role. The cluster Mo_6^{12+} contains 24 bonding electrons, the number appropriate for the 12 bonding orbitals available, whereas the cluster Nb_6^{14+} is the 16-electron species appropriate for the eight bonding orbitals. Both complexes are closed shell species, and the molybdenum complex $Mo_6Cl_8^{4+}$ in particular has the chemical stability appropriate for such a species. In $Mo_6X_8^{4+}$ ions in general, the eight X atoms are difficult to dislodge. However, each molybdenum atom has one additional ligand attached. This is located on the four-fold axis of the Mo_6 cluster, and can readily be varied, affording complexes such as $Mo_6X_8L_6^{4+}$, $Mo_6X_8YL_5^{3+}$, $Mo_6X_8Y_2L_4^{2+}$, etc. (L is a neutral ligand, Y a singly charged anion) whilst the $Mo_6X_8^{4+}$ core remains intact. By contrast, the niobium and tantalum species $M_6X_{12}^{2+}$ (which on this treatment contain only two-thirds as many metal–metal bonding electrons; in which the atoms X occupy doubly, not triply, bridging positions; and in which the overall positive charge is half that on the molybdenum complex) are much less chemically robust. Cluster species in general can be degraded by severe oxidation, pyrolysis, or base hydrolysis (cf. carboranes and borane anions).[12]

Since an octahedron has twelve edges and eight faces, the metal–metal bonding in both $Mo_6X_8^{4+}$ and $Nb_6X_{12}^{2+}$ can be described in terms of localized two- or three-centre bonds. The 12 cluster bonding electron pairs of $Mo_6Cl_8^{4+}$ could be allocated to 12 two-centre Mo–Mo bonds along the octahedral edges, and the eight bonding pairs of $Nb_6Cl_{12}^{2+}$ could be allocated to eight central three-centre bonds in the octahedral faces (contrast the situation that obtains for the cage bonding of the anion $B_6H_6^{2-}$, which has *seven* cage bonding pairs, a number inappropriate for a symmetrical two- or three-centre bond arrangement, p. 52). Note that the cluster described as held together by twelve edge bonds ($Mo_6Cl_8^{4+}$) has eight halogen atoms over the octahedral faces, whereas that described as having eight face bonds ($Nb_6Cl_{12}^{2+}$) has 12 halogen atoms over the octahedral edges. If the orbital contribution each halogen atom makes to the $M_6X_x^{n+}$ cluster bonding is resolved into radially and tangentially orientated components, and each halogen is assumed to contribute an electron pair in a radial orbital pointing towards the centre of the cluster, then the two types of cluster cation, $M_6X_8^{4+}$ and $M_6X_{12}^{2+}$, are effectively isoelectronic.[16] In short, the large difference between the numbers of cluster bonding electrons of the two types which is apparent when only metal–metal bonding is considered vanishes when certain metal–halogen interactions are also considered, when for both species there are $12+8 = 20$ pairs of cluster bonding electrons in 20 molecular orbitals, of symmetries $A_{1g}(2)$, A_{2u}, E_g, $T_{1u}(2)$, $T_{2g}(2)$, and T_{2u}.

The platinum(II) halides, $PtCl_2$ and $PtBr_2$, adopt hexameric structures Pt_6X_{12} formally analogous to the M_6X_{12} cluster shown in Fig. 7–8. However, they are held together effectively solely by the doubly bridging halogen atoms. That metal–metal bonding interactions are negligible is shown by the metal–metal distance in the chloride of 336 pm. This is not surprising, since the Pt_6^{12+} core these halides may formally be regarded as based on contains 48 electrons, the number appropriate to fill all the cluster antibonding, as well as the bonding, orbitals.

A few structural types found among metal carbonyl clusters are shown in Fig. 7–9. Again, molecular orbital treatments can be made along the lines indicated for $Mo_6Cl_8^{4+}$ and $Nb_6Cl_{12}^{2+}$, and again, where species have counterparts of the same symmetry among boron cage systems, differences between the bonding requirements of the two arise because of the involvement of d orbitals, and because of the presence of bridging groups in many of the transition metal species.

For example, further variations on the octahedral structural theme are provided by an interesting series of carbonyl complexes which include the rhodium compound $Rh_6(CO)_{16}$ which has six terminal $Rh(CO)_2$ units, and four triply bridging carbonyl groups over four alternate octahedral faces (see Fig. 7–9). This cluster compound contains a total of 86 valence shell electrons ($6 \times 9 = 54$ from the metal atoms, and $16 \times 2 = 32$ from the carbonyl groups) available for metal–metal and metal–carbonyl bonds, the same number as two other octahedral carbonyl species already described in Chapter 5 in connection with their bridging hydrogen and

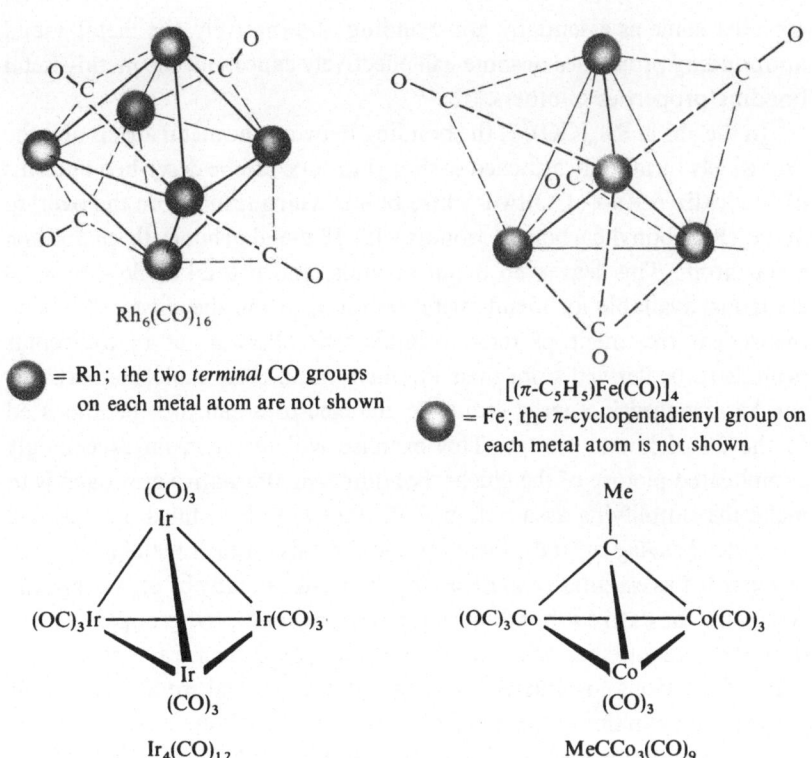

$Rh_6(CO)_{16}$

⬤ = Rh; the two *terminal* CO groups
    on each metal atom are not shown

$[(\pi\text{-}C_5H_5)Fe(CO)]_4$

⬤ = Fe; the π-cyclopentadienyl group on
    each metal atom is not shown

$Ir_4(CO)_{12}$

$MeCCo_3(CO)_9$

Fig. 7–9 Some transition metal carbonyl cluster compounds

six-coordinate carbon atoms, viz. the compounds $H_2Ru_6(CO)_{18}$ (Fig. 5–8d) and $Ru_6(CO)_{17}C$, in which the metal cluster serves as a cage about the carbide carbon (Fig. 5–25). The electron counts for these last two complexes involve eight electrons from each of their metal atoms, two electrons from each carbonyl group, one from each hydrogen atom and four from the carbide carbon.

That all 86 electrons in each of these complexes are not in strongly bonding orbitals is apparent from an orbital count. For example, consider the carbonyl hydride $H_2Ru_6(CO)_{18}$. All its carbonyl groups are terminally attached, and so each carbonyl group may be taken to contribute one orbital to the skeletal bonding, ignoring for the moment M→CO dative π-bonding effects, which will require the use of π* orbitals of the carbonyl groups. Each hydrogen atom contributes one orbital, and each ruthenium atom nine orbitals, to the cluster bonding, making an overall total of $18+2+9\times6 = 74$ orbitals. For the parent anion $Ru_6(CO)_{18}^{2-}$, only 72 orbitals are available, not more than half of which may be expected to be strongly bonding. Since 86 electrons require 43 orbitals to accommodate them (there are no unpaired electrons in these complexes) clearly all 43 orbitals cannot be strongly bonding, and it would not be unrealistic to consider some as essentially non-bonding; alternatively, the metal–metal antibonding properties of some will effectively cancel out the metal–metal bonding properties of others.

In the anion $Ru_6(CO)_{18}^{2-}$, the bonding between the metal atoms and the exclusively terminally attached carbonyl groups can be described in terms of 18 localized $Ru \leftarrow CO$ two-centre bonds which result from the overlap of the 18 carbonyl carbon sp orbitals with 18 metal orbitals, three on each metal atom. This leaves 36 metal orbitals, and a total of $86 - 36 = 50$ electrons, available for metal–metal bonding within the cluster. A group theoretical treatment of these orbitals will allow a set of molecular orbitals to be derived from their interactions; if these molecular orbitals can be arranged in energy sequence, the electrons can then be allocated to the 25 of lowest energy. This exercise will result in an exceedingly complicated picture of the cluster bonding. An alternative approach is to make the simplifying assumption that, since not all 25 filled orbitals will be cluster *bonding* orbitals, three atomic orbitals on each metal atom may be regarded as essentially *non-bonding* (they will actually be of appropriate symmetry for dative π-bonding to the terminal carbonyl groups). Allocation of a pair of electrons to each of these would then leave only seven pairs of electrons for cluster bonding orbitals derived from interactions between the remaining three atomic orbitals on each metal atom.

The picture this gives of the cluster bonding of $Ru_6(CO)_{18}^{2-}$ is reminiscent of that given for the anion $B_6H_6^{2-}$ (*cf.* Fig. 3–19). Each $Ru(CO)_3$ unit of $Ru_6(CO)_{18}^{2-}$, like each BH unit of the borane anion, effectively contributes two electrons, and three orbitals, to the cage bonding. A similar treatment can be applied to the related carbide $Ru_6(CO)_{17}C$, with the modifications that two 2-centre bonds to the doubly bridging carbonyl group replace two terminal Ru–CO links, and the central carbide carbon contributes four electrons for skeletal bonding. For $Rh_6(CO)_{16}$, four pairs of electrons per metal may be allocated to Rh–CO σ-bonds, and two pairs to cluster non-bonding (metal $\rightarrow$ carbonyl dative π-bonding) orbitals, leaving again three atomic orbitals on each metal, and altogether seven electron-pairs, for cluster bonding.[18]

This analogy between these octahedral carbonyl complexes and the borane anion $B_6H_6^{2-}$ can be taken a stage further by considering the compound $Fe_5(CO)_{15}C$, which also was described in Chapter 5 (Fig. 5–25, p. 146). This complex may formally be regarded as a derivative of the anion $Fe_5(CO)_{15}^{4-}$, the species with five terminal $Fe(CO)_3$ units at the corners of a square-based pyramid, which is left if one removes the carbon as C^{4+}. It was noted in Chapter 6 that *nido* structures are adopted by carboranes that are isoelectronic with borane anions $B_nH_n^{4-}$; their shapes are those appropriate for closed cage anions $B_{n+1}H_{n+1}^{2-}$, but with one cage site vacant. A similar structural relationship is seen between

$M_5(CO)_{15}^{4-}$ (as represented by $Fe_5(CO)_{15}C$, Fig. 5–25) and $M_6(CO)_{18}^{2-}$ (as represented by $H_2Ru_6(CO)_{18}$. Fig. 5–8d).

The remaining metal carbonyl cluster compounds shown in Fig. 7–9— $[(\pi\text{-}C_5H_5)Fe(CO)]_4$, $Ir_4(CO)_{12}$, and $MeCCo_3(CO)_9$—illustrate structures based on a tetrahedron or triangle of metal atoms. The bonding in the last two, which have no bridging carbonyl groups, can be described in terms of localized two-centre electron pair bonds along the edges of the Ir_4 tetrahedron or Co_3C trigonal pyramid, with again three pairs of electrons on each metal atom in what are effectively non-bonding orbitals as far as metal–metal interactions are concerned. Treating the iron complex as containing an $[Fe(CO)]_4^{4+}$ core to which four $C_5H_5^-$ ions donate six electrons apiece, leaves a total of 36 electrons for core bonding, enough for six two-centre metal–metal tetrahedral edge bonds and also for 12 two-centre metal carbon bonds to the triply bridging carbonyl groups.

To summarize: in this section we have briefly surveyed the numbers of electrons available for cluster bonding in some transition metal systems based on octahedral or tetrahedral clusters of metal atoms. Most have turned out to contain enough electrons to allow their bonding to be described in terms of two-centre bonds, and so are not included by our definition of 'electron deficient'. However, some points of similarity between transition metal species and cage borane anions have emerged. Though the greater number of bonding orbitals available in the transition metal species leads to a greater variety of arrangements of substituents, bridging as well as terminal, about the cluster, and also to more complicated sets of interactions between atomic orbitals within the clusters, nevertheless the underlying bonding requirements of particular clusters are reflected by such parallel relationships as those between $Ru_6(CO)_{18}^{2-}$ and $Fe_5(CO)_{15}^{4-}$ and between $B_6H_6^{2-}$ and $B_5H_5^{4-}$. There are clearly advantages in treating cluster species in general as a class, rather than arbitrarily distinguishing those that contain relatively few bonding electrons from those that contain enough to allocate two to each point of contact between cluster atoms. If distinctions between structurally related cluster species are to be made on the basis of the number of valence shell electrons they contain, then the important criteria are (a) whether all the cluster bonding orbitals are filled (oxidizing properties are to be expected if they are not), and (b) whether extra electrons are accommodated in essentially non-bonding, or antibonding, cluster orbitals (as this will influence metal–metal bond distances, cf. $Nb_6Cl_{12}^{2+}$ and Pt_6Cl_{12}).

7–4 Transition metal-hydrocarbon π-complexes

In the discussion of metallo-carboranes in Chapter 6, it was noted that the metal environment provided by a symmetrically coordinating dicarbollide ligand $C_2B_9H_{11}^{2-}$ was like that in π-cyclopentadienyl complexes. The similarity is underlined by the existence of such mixed sandwich species as the iron(III) complex $(\pi\text{-}C_5H_5)Fe[(3)\text{-}1,2\text{-}C_2B_9H_{11}]$ (Fig. 6–7), in which the metal is effectively equidistant (*ca.* 205 pm) from not only all five carbon atoms of the cyclopentadienyl ring but also the two carbon atoms and three borons of the open pentagonal face of the dicarbollide residue. There is clearly no structural justification for invoking more than one type of bonding between the metal and the two ligands in this complex, in which the 10 atoms in the metal's coordination sphere cannot be bound by localized two-centre M–C and M–B bonds, there being at most 18 electrons likely to be involved in the valence shell of the metal.

This shortage of electrons is not just a peculiarity of mixed sandwich species. Many coordination complexes formed by transition metals with hydrocarbon ligands are electron deficient in that they possess too few bonding electrons to allow a pair to be allocated to each of their metal–carbon links. In ferrocene, $(\pi\text{-}C_5H_5)_2Fe$, and dibenzenechromium, $(C_6H_6)_2Cr$, for example, there are 10 and 12 carbon atoms respectively within a covalent bonding distance of, and equidistant from, the metal atom, which in each case has a total of 18 electrons in its valence shell. However, it is not customary to classify such complexes as electron deficient or otherwise, because treatments of their bonding in terms of localized two- or three-centre bonds are rarely helpful, and so the question of whether a particular compound happens to contain enough electrons to allow a two-centre bond description becomes a relatively unimportant technicality. The normal approach is to use a partial molecular orbital treatment which takes account of the number of electrons, and the number and type of the atomic orbitals, that a particular ligand may be reckoned to contribute towards metal–ligand bonding. Details are to be found in texts directly concerned with these important complexes.[17] It is worth illustrating the approach here by reference to the manner in which a regular pentagonal cyclopentadienide anion $C_5H_5^-$ can coordinate to a metal ion located over its centre on the five-fold axis (see Fig. 7–10), because the pentagonal pyramidal arrangement of the six atoms involved is a structural feature common to many higher boranes, borane anions, and carboranes as well as metallo-carboranes.

    The ligand orbitals of the cyclopentadienide anion available for metal–ligand bonding are shown in Fig. 7–10. Only those ligand orbitals pointing in the general direction of the metal, the five p_z atomic orbitals of the carbon atoms, need be considered. These interact under the local C_{5v} symmetry to generate the set of five ligand π-orbitals shown. Lowest in energy is orbital (a), of A_1 symmetry. Next is a degenerate pair, (b) and (c), of E_1 symmetry. Highest in energy is another degenerate pair of orbitals, (d) and (e), of E_2 symmetry. Orbitals (a), (b), and (c) are those which would

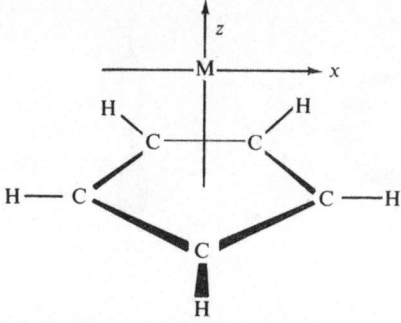

The pentagonal pyramidal (C_{5v}) structure
and axes of the π-cyclopentadienyl-metal unit

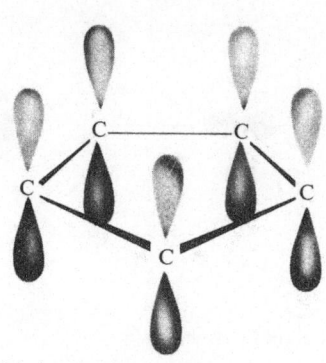

(a) Appropriate symmetry (A_1) to
interact with metal d_{z^2}, s and p_z orbitals

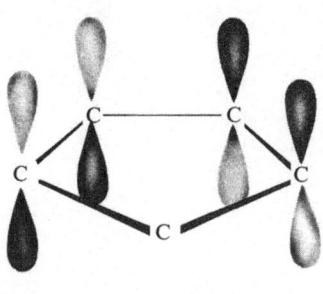

(b) Appropriate symmetry (E_1) to
interact with metal d_{xz} and p_x orbitals

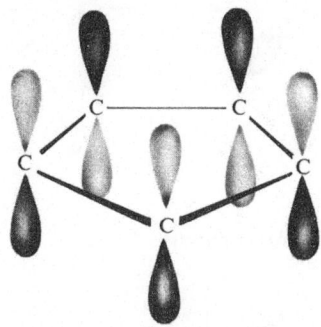

(c) Appropriate symmetry (E_1) to
interact with metal d_{yz} and p_y orbitals

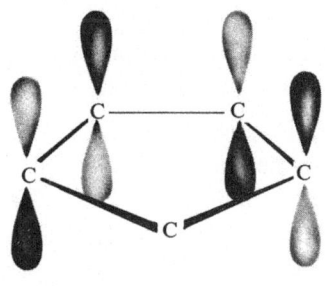

(d) Appropriate symmetry (E_2) to
interact feebly with metal d_{xy} orbital

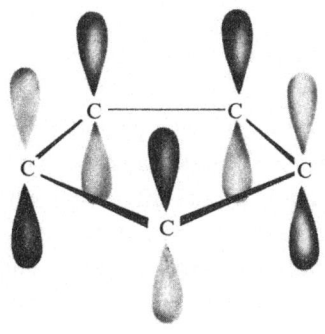

(e) Appropriate symmetry (E_2) to
interact feebly with metal $d_{x^2-y^2}$ orbital

Fig. 7–10   Ligand orbitals available for π-cyclopentadientl-metal bonding

Orbitals of A_1 symmetry:

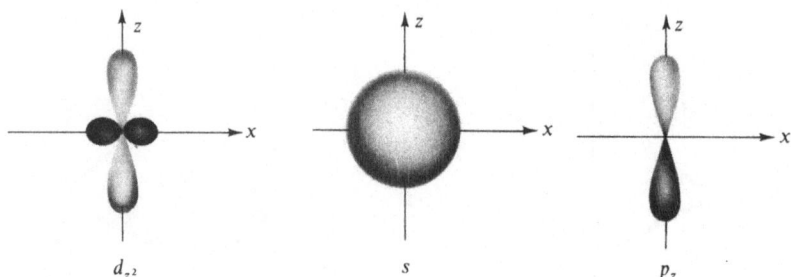

Orbitals of E_1 symmetry:

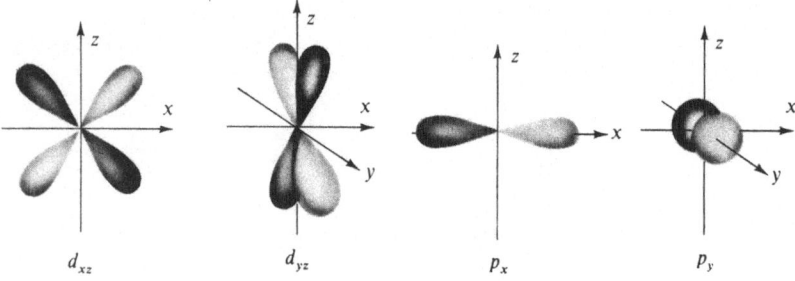

Orbitals of E_2 symmetry:

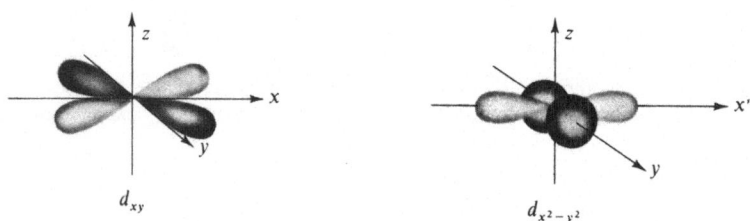

Fig. 7–11   Metal orbitals available for π-cyclopentadienyl-metal bonding

be filled in the cyclopentadienide anion $C_5H_5^-$, which has six π-bonding electrons. Cyclopentadienyl complexes may be regarded as derived from this anion by its interaction with a metal cation on the five-fold axis. Ligand $\rightarrow$ metal dative bonding requires the use of metal orbitals (Fig. 7–11) of symmetry A_1 (d_{z^2}, s, and p_z in order of increasing energy) and E_1 (d_{xz}, d_{yz}, and p_x, p_y). Feeble metal $\rightarrow$ ligand back coordination is possible for the orbitals of E_2 symmetry, from filled metal d_{xy} or $d_{x^2-y^2}$ orbitals. The precise energy sequence of the molecular orbitals that result from such interactions between ligand and metal orbitals will depend on the metal and on what other ligands are present. Similar treatments of the metal–ligand interactions in other π-complexes can be made, the complexity of these interactions varying with the number of p-orbitals contributed by the ligand. The great virtue of the method is that it allows one to deduce in what manner the bonding in the ligand (which can generally also be obtained in the uncoordinated form) will be modified by the presence of the metal, which by draining electronic charge from ligand π-bonding orbitals and releasing electronic charge into ligand π-antibonding orbitals will reduce the bond order in the ligand and therefore modify its chemical reactivity.

An analogous treatment of dicarbollide–metal interactions is possible, making use of five p-orbitals of the two carbon and three boron atoms surrounding the pentagonal face of the ligand. The five p-orbitals in question will be five tangential to the pseudo spherical surface of the ligand cage. Extension of the method to a description of the bonding between one boron atom and its five neighbours in an icosahedral borane anion or carborane is, however, less helpful than treatments which consider the whole cage. But it does provide a useful alternative treatment in other cases. For example, the pentagonal bipyramidal anion $B_7H_7^{2-}$ can be regarded as a hypothetical pentagonal anion $B_5H_5^{6-}$ (isoelectronic with $C_5H_5^{-}$) interacting with a pair of suitably orientated BH^{2+} cations, one above, the other below, the centre of the pentagon. If the boron atoms of the BH^{2+} cations are regarded as sp hybridized, with one sp hybrid orbital used for the terminal BH bond, there remain one sp hybrid orbital pointing towards the centre of the pentagon, and two p orbitals at right angles to this, available for bonding the apical borons to the five equatorial borons. These have the appropriate symmetries (A_1 and E_1 respectively) to interact with the filled π-orbitals of the $B_5H_5^{6-}$ species.

7–5   Electron deficient sigma bonding in electron rich species

All the electron deficient compounds discussed in this book have contained an atom or atoms that supplied fewer valence electrons than low energy atomic orbitals for bonding. In order to make use of all the available orbitals they have adopted structures in which a small number of strongly bonding molecular orbitals resulted from interactions between these atomic orbitals. They have left unused a greater number of essentially non- or anti-bonding molecular orbitals that also resulted from the atomic orbital interactions dictated by the molecular symmetry. Electrons added to these orbitals would not have increased the stability of the molecular structure. There are circumstances, however, in the chemistry of relatively electron rich species, where non- or weakly anti-bonding orbitals are occupied, and molecules may be held together by fewer than two bonding electrons per two-centre link despite an apparent abundance of electrons for bonding. These are derivatives of elements that suffer from a deficiency of low energy atomic orbitals rather than of electrons.

Typical of such substances are certain halides of elements from the right-hand side of the periodic table, like sulphur tetrafluoride, xenon difluoride, or iodine trichloride which formally, on a two-centre electron-pair bonded picture, require more than eight electrons to be accommodated in the valence shell of the central atom. A common structural feature is a more or less linear XMX arrangement of two halogen atoms X about the central atom M, which may also have one or more other atoms attached. The M–X distances in these are longer than might be expected for M–X single bonds.

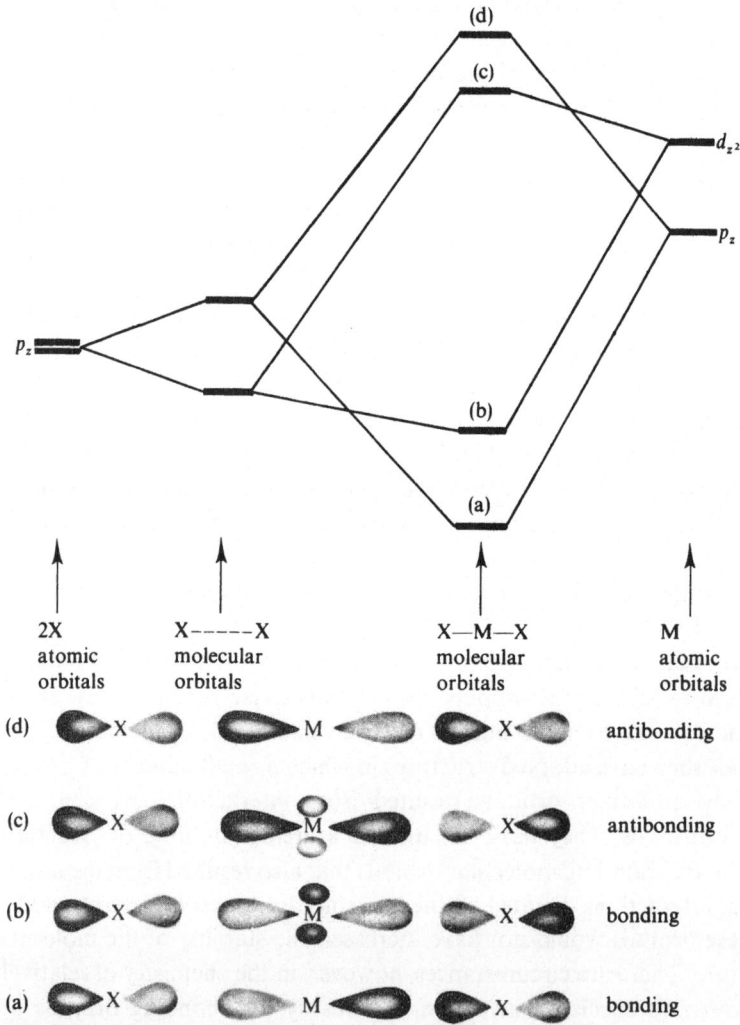

Fig. 7–12 Orbital correlation diagram for a linear XMX species in which the central element M contributes two orbitals for bonding

The bonding in these compounds can be understood by considering alternative representations depending on the number of orbitals effectively contributed by the central atom M. In both representations, the atom M is considered to supply two electrons, and each atom X one electron, towards XMX bonding. If M has available two low energy atomic orbitals, for example, a p_z and a d_{z^2} orbital, pointing to the two halogen atoms X, then interactions between these two orbitals and an atomic orbital of each halogen will generate a set of four orbitals, two bonding, two antibonding, as shown in Fig. 7–12. The four electrons associated with these orbitals may then be allocated to the two bonding orbitals, generating effectively two two-centre electron pair bonds, that is, the M–X bond order is one.

A different situation obtains if the d-orbital of the element M is of very high energy, leaving only the p orbital of suitable energy to interact significantly with the p orbitals of the halogen atoms X. The orbital correlation diagram is then as shown in Fig. 7–13. The lowest energy three-centre orbital formed is that corresponding to the open three-centre bond discussed in Chapter 3 (Fig. 3–2). Next in energy is a non-bonding orbital, formally weakly X–X bonding, to which the p-orbital of M makes no contribution because the wave function for this p-orbital changes sign passing through M along the X–M–X axis. The third and highest energy three-centre orbital is antibonding with respect to both M–X and X–X interactions. Of the four electrons to be accommodated in these orbitals, two go into the bonding orbital, and the other two into the non-bonding orbital. On this treatment, the M–X bond order is 0·5, and bearing in mind from which atoms the electrons originated (but ignoring any difference in electronegativity between M and X), the charge distribution is $X(-\frac{1}{2})\cdots$ $M(+1)\cdots X(-\frac{1}{2})$.

In actual compounds, an M–X bond order between 0·5 and 1·0 is likely, since the bonding character of orbital (b) in Figs 7–12 and 7–13 is a function of the p_z–d_{z^2} energy difference; a gradual decrease in energy of the d_{z^2} orbital of M (Fig. 7–13) will cause a gradual decrease in energy of the three-centre orbital (b), which will change in character from essentially non-bonding to bonding, as in Fig. 7–12.

One compound in which the M–X bond order appears to be appreciably less than unity is xenon difluoride XeF_2, in which the Xe–F interatomic distance is 200 pm. This is significantly longer than that (184 pm) in its bis-antimony pentafluoride adduct $XeF_2.2SbF_5$, the structure of which corresponds to the ionic formulation $XeF^+Sb_2F_{11}^-$, although there is weak interaction between the xenon and one fluorine of the anion. These xenon–fluorine distances are intelligible in terms of a change of bond order from about 0·5 in xenon difluoride itself, to virtually a full single bond in the cation Xe–F$^+$, the bonding in which requires the use of no xenon d orbitals. The participation of xenon d orbitals may be justified

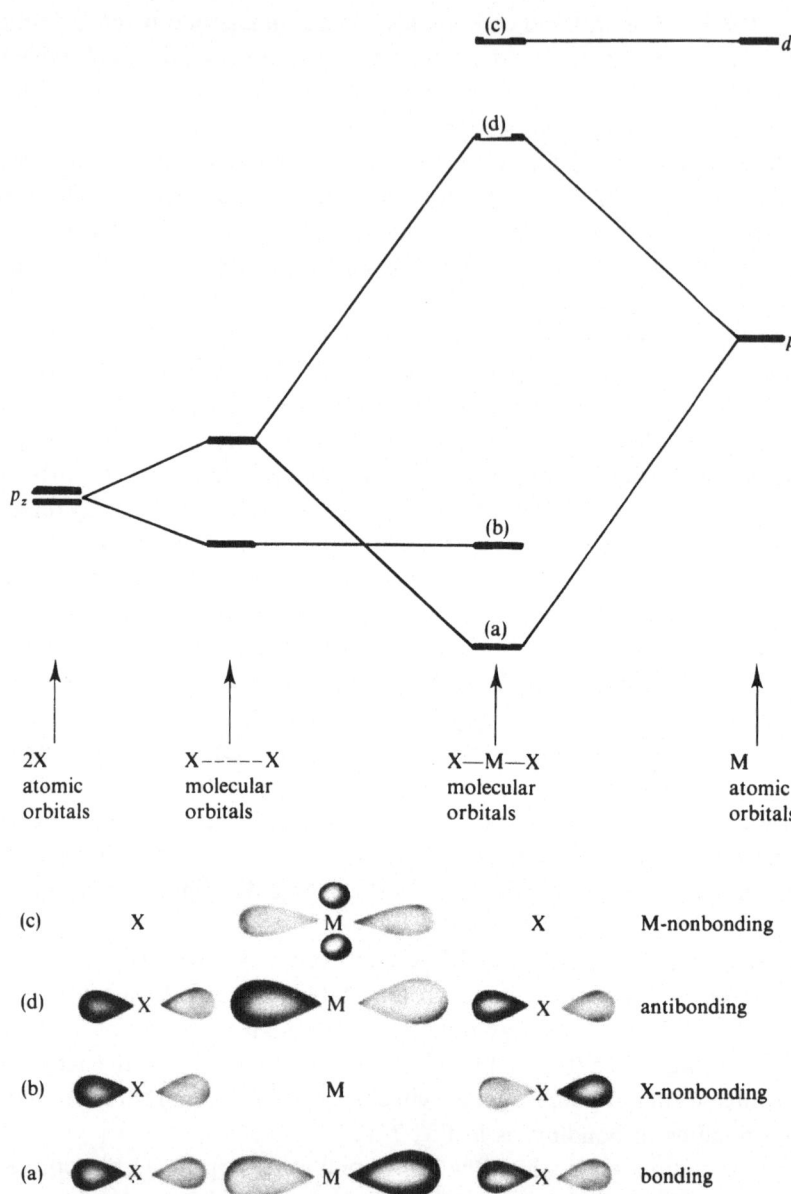

Fig. 7–13 Orbital correlation diagram for a linear XMX species in which the central element M contributes one orbital for bonding

in the higher fluorides, which have Xe–F distances of 195 pm (XeF$_4$) and 189 pm (XeF$_6$).

Similar arguments may be applied to other halides containing essentially linear X–M–X groupings for which two-centre electron pair bond schemes would require the use of unrealistically high energy orbitals of the central element M. For example, they allow a rationalization of the greater length of two of the three MX bonds in the interhalogens MX_3 (for example, ClF_3 has two bonds of 170 pm, and one of 160 pm), or of two of the four SF bonds in SF_4 (which has bond distances of 165 and 155 pm). Three three-centre bonds could account for the regular octahedral structures found for some MX_6 species containing seven electron pairs, like the anion $SeCl_6^{2-}$; of the remaining four 'bonding' pairs of electrons, one would be accommodated in the s orbital of the central element, and the other three in the three non-bonding orbitals (type (b) of Fig. 7–13) essentially localized on the substituent halogen atoms.

The bonding in all these electron rich, orbital deficient species probably is intermediate between the types shown in Figs 7–12 and 7–13. Nevertheless, their low M–X bond orders show that the concept of three-centre sigma bonding orbitals has much wider application than solely to electron deficient compounds, and can profitably take its place alongside multi-centre pi-bonding which, as a supplement to two-centre sigma bonding, has been a long familiar concept in the chemistry of unsaturated systems.

References

1. Field, F. H., and M. S. B. Munson, *J. Amer. Chem. Soc.*, 1965, **87**, 3289.
2. Olah, G. A., G. Klopman, and R. H. Schlosberg, *J. Amer. Chem. Soc.*, 1969, **91**, 3261.
3. Worstall, R. A., *J. Amer. Chem. Soc.*, 1899, **21**, 245.
4. Clark, D. T., and D. M. J. Lilley, *Chem. Comm.*, 1970, 549.
5. Clark, D. T., and D. M. J. Lilley, *Chem. Comm.*, 1970, 1042.
6. Davis, B., D. H. Williams, and A. N. Y. Yeo, *J. Chem. Soc. (B)*, 1970, 81.
7. Olah, G. A., and J. Lukas, *J. Amer. Chem. Soc.*, 1967, **89**, 4739.
8. Olah, G. A., and J. M . Bollinger, *J. Amer. Chem. Soc.*, 1967, **89**, 4744.
9. Bethell, D., and V. Gold, The structure of carbonium ions, *Quart. Rev.*, 1958, **12**, 173–203.
10. See for example, Sargent, G. D., Bridged, non-classical carbonium ions, *Quart. Rev.*, 1966, **20**, 301–371 for early work.
11. Olah, G. A., A. M. White, J. R. DeMember, A. Commeyras, and C. Y. Lui, *J. Amer. Chem. Soc.*, 1970, **92**, 4627.
12. Cotton, F. A., Transition metal compounds containing clusters of metal atoms, *Quart. Rev. (London)*, 1966, **20**, 389–401.
13. Penfold, B. R., Stereochemistry of metal cluster compounds, *Perspectives in Struct. Chem.*, 1968, **2**, 71–149; Chini, P., The closed metal carbonyl clusters, *Inorg. Chim. Acta Rev.*, 1968, **2**, 31–51.
14. Baird, M. C., Metal–metal bonds in transition metal compounds, *Progress in Inorg. Chem.*, 1968, **9**, 1–159; Abel, E. W., and F. G. A. Stone, The chemistry of transition-metal carbonyls; structural considerations, *Quart. Rev. (London)*, 1969, **23**, 325–371.
15. Cotton, F. A., and T. E. Haas, *Inorg. Chem.*, 1964, **3**, 10.
16. Kettle, S. F. A., *Theoretica Chim. Acta*, 1965, **3**, 211.
17. Green, M. L. H., The transition elements, Vol. II of *Organometallic Compounds* by G. E. Coates, M. L. H. Green, and K. Wade, 3rd Edn., Methuen, London, 1968; Coates, G. E., M. L. H. Green, P. Powell, and K. Wade, *Principles of Organometallic Chemistry*, Methuen, London, 1968.
18. Wade, K., *Chem. Comm.*, 1971, 792.

Problems

7-1 Explain how ^1H n.m.r. spectroscopy can be used to elucidate the structures of carbonium ions.

7-2 Explain what carbonium ions will be generated in a mixture of fluorosulphuric acid and antimony pentafluoride by the following: (a) methane; (b) 2-chloro-2-methylpropane; (c) 1-chloro-2-methylpropane; (d) 2-chloro-2,3-dimethylbutane; (e) 3-ethylpentane.

Describe the ^1H n.m.r. spectroscopic absorptions these carbonium ions will give rise to.

7-3 The compound $Ru_5(CO)_7(\pi\text{-}C_5H_5)_4$ has been prepared. Assuming that its four cyclopentadienyl groups may be regarded as terminally coordinated $C_5H_5^-$ ligands acting as six-electron donors, suggest a likely structure for this compound.

7-4 Explain why the metal–metal distance in the platinum dichloride hexamer is much greater than that in compounds based on the $Nb_6Cl_{12}^{2+}$ cluster.

7-5 If the bridging hydrogens of hexaborane(10) are arbitrarily removed as protons, the residual anion $B_6H_6^{4-}$ may be regarded as formed by the interaction of a pentagonal $B_5H_5^{6-}$ anion with a cation BH^{2+}. Construct an orbital correlation diagram to show what bonding orbitals may be expected to result from such an interaction.

7-6 Explain how the concept of the three-centre electron pair bond may be helpful in a discussion of the bonding in systems where there appear to be ample electrons for two-centre electron pair bonds.

7-7 Discuss critically the appropriateness of the term 'electron deficient' as applied to substances containing too few pairs of bonding electrons to allow all their covalent bonds to be described in terms of two-centre electron pair bonds.

Answers to problems

Chapter 2

2-2 v_8, b_{2u}; $v_{13,14}$, b_{3u}; $v_{16,17,18}$, b_{1u}

2-3 Diagram very similar to Fig. 2–12

2-4 $B_2H_6^{2+}$ might have such a structure, not B_2H_6

2-5 (a, b) ^1H, 1111 quartet; ^{11}B, 1331 quartet

(c) ^1H, 1111 quartet, 1234321 septet; ^{11}B 1331 quartet of doublets

(d) ^1H, as for (c), but intensities 4:1, not 6:1; ^{11}B 121 triplet of doublets

Chapter 3

3-1 a, 0; b, $+\frac{1}{3}, -\frac{1}{6}$; c,f, $0, +\frac{1}{2}, -\frac{1}{2}$; d,e, 1, $-\frac{1}{2}$; f, $-\frac{1}{6}, +\frac{1}{6}$

(b) a, -1; b, $-\frac{1}{2}$; c, 0, $-\frac{1}{2}$; d, $-\frac{2}{3}, -\frac{1}{6}$

(c) $-1, 0, +\frac{1}{2}$

(d) B(1, 3, 6, 9), 0; B(2, 4), $-\frac{2}{3}$; B(5, 7, 8, 10), $+\frac{1}{3}$

3-3 $(2010)B_2H_4$, $(1020)B_2H_3^-$, $(1012)B_2H_5^-$, $(1004)B_2H_7^-$, $(0030)B_2H_2^{2-}$, $(0022)B_2H_4^{2-}$, $(0014)B_2H_6^{2-}$, etc.

3-4 (1301) and $(2210)B_4H_6$, $(4004)B_4H_{12}$, (1401) and $(2301)B_5H_7$, and (5013) and $(4104)B_5H_{13}$ appear possible

3-6 (2402) and $(3311)B_6H_{10}$ are theoretically possible, as are (3303), (4212), (5121), and $(6030)B_6H_{12}$. However, B_6H_{12} should adopt an *arachno*-structure based on a dodecahedron, not a *nido*-structure based on a pentagonal bipyramid (see Chapter 6).

Chapter 4

4-1 (a) $x = 3p$, $y = p$, $z = \frac{1}{2}(4p+q)$

(b) B_2H_6; (i) 574, (ii) 20·7: B_4H_{10}; 1152, 21·6: B_5H_9; 1427, 22·6: B_5H_{11}; 1453, 22·1: B_6H_{10}; 1718, 23·0

4-2 Me_3N, 117; Me_3As, 121; Me_2S, 120; Me_2Se, 109; CO, 95; Me_3P (Δ H), 57 kJ/mole

4-3 (a) X = F, Δ H = 672; Cl, 288; Br, 228; I, 60 kJ/mole

(b) X = F, Δ H = 3924; Cl, 1392; Br, 846; I, 114 kJ/mole

4-4 (a) Trap B_4H_{10} at $-126°$, B_2H_6 at $-196°$; Töpler for H_2

(b) Trap $B_{10}H_{14}$ at -20 to $-78°$, B_5H_9 at -95 to $-126°$, B_2H_6 at $-196°$

(c) $B_2H_6 + Me_3N \rightarrow Me_3N,BH_3$ (involatile) $+\frac{1}{2}B_2H_6$

4-8 Me_2NH,BH_3; $Me_2NB_2H_5$; Me_2NBH_2; $(Me_2N)_2BH$; $B(NMe_2)_3$

4-15 1,2-Tetramethylenediborane Hydrolysis → $(HO)_2B(CH_2)_4B(OH)_2$

B = $Me_3N,BH_2(CH_2)_4BH_2$, NMe_3

4-16 A = B_6H_{10} B = B_5H_9
C = B_4H_{10}

4-17 A = $Me_2BH_2BH_2$
B = $(EtNBH)_3$
C = *trans*- $(Me_2BN:CHMe)_2$
D = *cis*- $(Me_2BN:CHMe)_2$

Chapter 5

5-3 ΔH_a = 4618 kJ/mole
E(AlHB) = 395 kJ/mole

5-4 (a) $Me_2AlPh_2AlMe_2$
(b) reactants plus $Me_2AlBrClAlMe_2$
(c) $Al(BH_4)_3 + Me_nB_2H_{6-n}$
(d) $Al(OH)_3$ + methane (e,f) $Me_2AlMe(OBu^t)AlMe_2$

5-7 B, 97; Al, 78; Ga, 34

Chapter 6

6-1 1,2–; 1,7–; 2,3–; and 2,4–$C_2B_5H_7$

6-2 1,2–; 1,6–; 2,3–; and 2,4–Me_2-1,6–$C_2B_4H_4$ 1,2–; 1,3–; 1,4–; 1,6–; 3,4–; 3,5–; and 4,6–Me_2-1,2–$C_2B_4H_4$

6-3 *cf* carboranes, replacing CH by NH^+

6-4 $CB_4H_5^-$, $CB_8H_9^-$, $MeAlC_2B_4H_6$, $[(C_2B_4H_6)_2Fe(II)]^{2-}$ *closo*-; $CB_4H_5^{3-}$, *nido*-; $C_4B_2H_6^{2-}$, *arachno*-; CBH_6^-, *cf* ethane

6-6 1,3–Me_2-1,2–$C_2B_3H_3$

Chapter 7

7-2 (a) to (c), Me_3C^+;
(d) $Me_2HCCMe_2^+$;
(e) Et_3C^+

7-3 *cf* $Fe_5(CO)_{15}C$ (Fig. 5–25), with apical $Ru(CO)_3$, and 4 basal $(C_5H_5)Ru$ units bridged by 4 carbonyl groups

Index